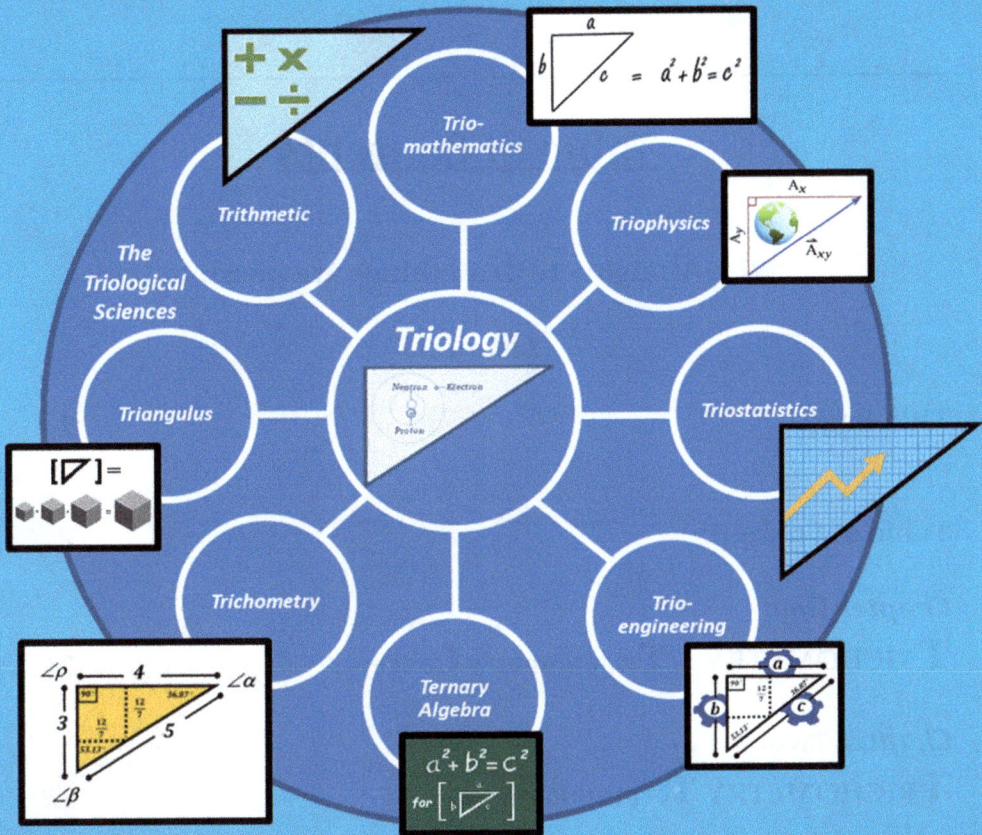

TRICHOMETRY ™ ©

The Study of the Geometrics
Of the 3-4-5-6 Upright Right
Triangle in Cartesian Coordinates

By

James E. Osler II, Ed.D.

Table of Contents

"With Almighty GOD all things are possible."

Luke 1:37

Library of Congress Cataloging–in–Publication Data

Osler II, James E.

TRICHOMETRY ™ © *The Study of the Geometrics of the 3-4-5-6 Golden Upright Right Triangle in Cartesian Coordinates.* 1st edition.

ISBN: 978–0–9826748–17–5

1. English Language–Mathematics. 2. Research, 3. Analysis, and 4. Instructional Design

Published in the United States.

ISBN: 978–0–9826748–17–5

Dedication and Special Acknowledgements

Dedication

I send a great and appreciative "Thank You" to Almighty GOD for his continual blessings and for his son my Lord and Savior Jesus Christ, through whom all things are possible. This book is designed to bless all who read it and provide them with greater insight and understanding. Oh Lord, be glorified in all that I do.

Special Acknowledgements

I also send a very special "Thank You" to my family. Your support has greatly aided me in seeing this project completed to the very end. You are a blessing. Thank you and I love you all.

About the Author

A native of Durham, N.C., James Edward Osler II is an experienced artist, eduscientist, entrepreneur, researcher, statistician, teacher, and technologist. He accepted his call into the ministry in 2014. He is an active member of North East Baptist Church where he serves in many capacities. He is also a tenured full professor in the School of Education at North Carolina Central University (NCCU). Osler has authored an abundance of influential refereed international journal articles, books, and papers. His current research agenda primarily consists of the three main topics they are: (1.) "Applied Educational Science"; (2.) "Innovative Measurement Methodologies"; and (3.) "Ergonomic E-Learning Engineering". He is licensed in the following areas: The Gospel Ministry (Preaching), Art Education (Secondary), and Instructional Technology Specialist: Computers (Graduate Level). He is a certified Microsoft Innovative Educator (MIE) and is also a certified Entrepreneurial Mindset Facilitator with additional certifications in Digital Literacy, Research, and Visual Programming.

Dr. Osler has served on multiple international journal editorial referee review boards and has created and authored several graduate-level certificate and degree programs. North Carolina State University (NCSU) has recognized him as one of its 100 Most Influential Black Alumni. He has received four of the highest and most respected honors at NCCU: "The Employee Recognition Award for Outstanding Service"; "The University Award for Teaching Excellence", "The Faculty Senate Award for Outstanding Scholarly Achievement"; and the first "Chancellor's Award for Innovation". He is also the recipient of three national awards for both his research and his service to the community: An Award from The National Council for American Executives; The IGI Global Outstanding Scholars Award; and The Church of GOD in Christ—Black History Achievement Award.

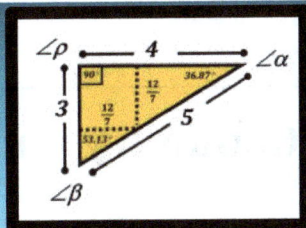

Give instruction to a wise man, and he will be yet wiser: teach a just man, and he will increase in learning.

Proverbs 9: 9

Defining Trichometry and a Detailed List of Geometric Concepts Relevant to Trichometry

Trichometry (pronounced "[Try · kom · et · tree]") is defined as: "The Trichotomous Metrics (or "Measurement") of the 3-4-5-6 Golden Upright Right Triangle (commonly referred to by the acronym of GURT)".

Traditional Geometric Concepts

The geometric mathematical review that follows is a detailed and comprehensive guide through classical geometric mathematical concepts. It is designed to provide the reader with an enriching source of information as the starting point to aid in the comprehension of "Trichometric Concepts". Understanding the mathematics of Geometry aids in forming a good foundation for the later comprehension of Trichometric calculations. This Guide can be viewed as both a starting point for initial study and as a reference tool that can be referred to at any time.

Geometry and Geometric Terminology

The term "Geometry" literally means "earth measure" it is the branch of mathematics concerned with the deduction of the properties, measurement, and the relationships of points, lines, angles, and shapes. Geometry defines these figures, their conditions, and their assumed properties in space.

Examples of Geometric figures related to the circle shape:

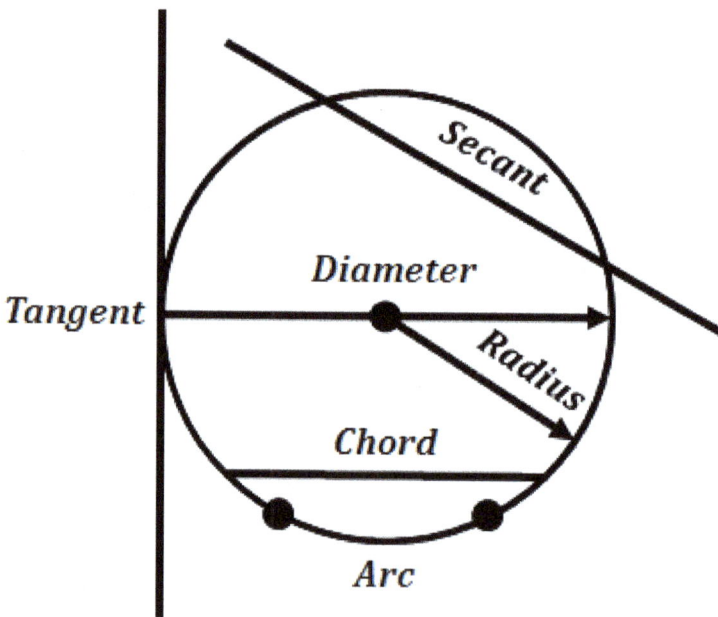

Classical Geometrical Definitions

Geometry uses three main areas to define and describe different figures they are: Postulates, Proofs, and Theorems.

Each term is defined as follows:

Postulate

The term "Postulate" in the mathematics of Geometry is a self-evident truth that requires no proof.

Proof

The term "Proof" in the mathematics of Geometry is a rational or logical sequence of steps usually provided as a series of statements (sometimes with or without illustrations) that lead to a valid conclusion.

Theorem

The term "Theorem" in the mathematics of Geometry is a theoretical formula (a sequence of steps that use mathematical logic), a proposition (a statement in which something is affirmed or denied), or a statement embodying something to be proved from other propositions or formulas.

To mathematical concepts crucial to developing an in–depth understanding of Geometry are Degree and Ratio.

They are defined as follows:

Degree

The definition of the term "Degree" in the mathematics of Geometry is the 360th part of a complete angle or turn (as related to a circle which has a total of 360 degrees = 360°). Each complete angle degree unit is represented by the sign "°" which immediately follows a number between 1 and 360. For example: "Thirty degrees" is written: "30°". Triangles have a total of 180 degrees when all three of the angles contained within the shape are added together.

Ratio

In mathematics a "Ratio" is the terminology used to define a relation between two similar magnitudes with respect to the number of times the first contains the second. For example – The ratio of 3 to 1, can be written in the following ways as common way of notation: 3:1 or as 3/1.

Other Common Definitions Relevant to Geometry

Angle

An "Angle" is the space within two lines that originate and diverge from a common point. It is sometimes written in Geometry in the following manner: □ . An illustration of a basic angel is:

An angle can be measured in either degrees or radians. A circle has a total of 360 degrees or 2π radians. Note:

$$\frac{360°}{360°} = \frac{2\pi}{360°} = \frac{2\pi}{360°} \div \frac{2}{2} = \frac{\pi}{180°} = \pi/180° = 1°$$

Thus, 1 degree = $\pi/180$ radians. Standard angles have the following measures:

$1° = \pi/180$ radians
$30° = \pi/6$ radians
$45° = \pi/4$ radians
$60° = \pi/3$ radians
$90° = \pi/2$ radians
$180° = \pi$ radians
$360° = 2\pi$ radians

Arc

An "Arc" is a segment of a line on the circumference of a circle. It is illustrated as follows:

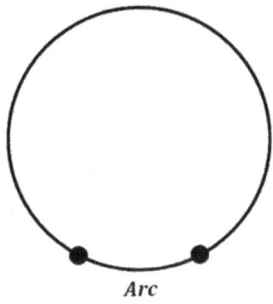

Arc

Area

The term "Area" applies to the quantitative measure of a plane or curved surface. The dark area of the circle that follows is an illustration:

Circumference

The "Circumference" of a circle (or circular area) is its border or outer boundary also called its perimeter. It is illustrated as follows:

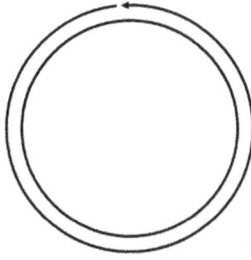

Chord

A "Chord" is a line segment between two points on a given curve. It is illustrated as follows:

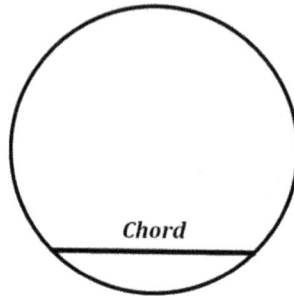

Chord

Circle

A "Circle" is a closed plane curve that creates a balanced shape consisting of all points that are an equal distance from a point within the shape called the "center". It is illustrated as follows:

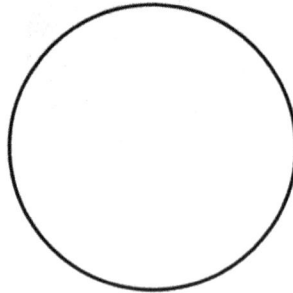

Cone

A Cone is a three-coordinate Geometric shape that is essentially a pyramid with a circular cross–section. A right cone is an uprightly positioned cone with its vertex (the point where all of its lines meet) above the center of its base (the circular section of the figure). In mathematics, the term "Cone" often means "right cone" (indicating that the cone is positioned in an upright fashion).

The following illustration displays a Cone:

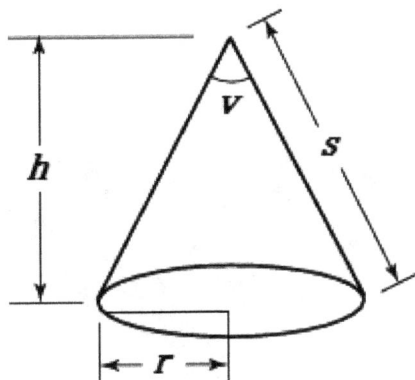

Where,

v = Vertex
h = Height
s = Side
r = Radius

The following illustration is a three-coordinate wireframe map of an upright Cone that illustrates the sides of the Cone meeting at its Vertex. The wireframe also illustrates how an upright Cone is essentially a smooth infinitely multi–sided pyramid with a circular cross–section (bottom):

A three-coordinate illustration of a wireframe metal sculpture of two symmetrical diametrically opposed cones (that are positioned downward and upright) connected at their respective vertexes is illustrated as follows:

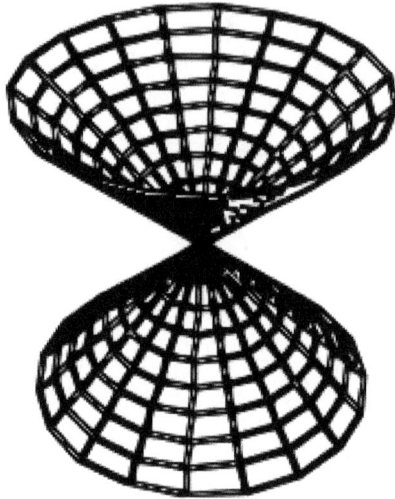

Conic Sections

In the mathematics of Geometry, a Conic Section is created when any curve produced by the intersection (or slicing) of a plane with an upright and/or inverted circular cone. Depending on the angle of the plane when it intersects or slices the cone, the intersection can create one of the following figures: a Circle, an Ellipse, a Hyperbola, or a Parabola. Each of these four aforementioned figures is considered a Conic Section.

The following two illustrations depict the Conic Sections that are created when a plane intersects (or slices) a cone:

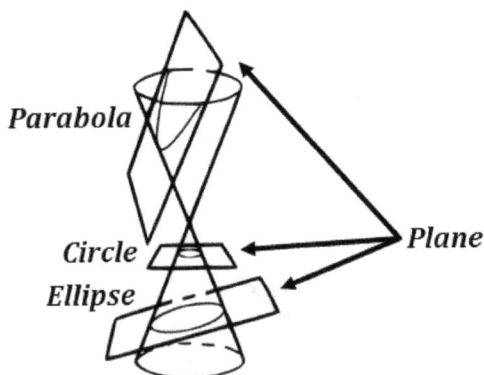

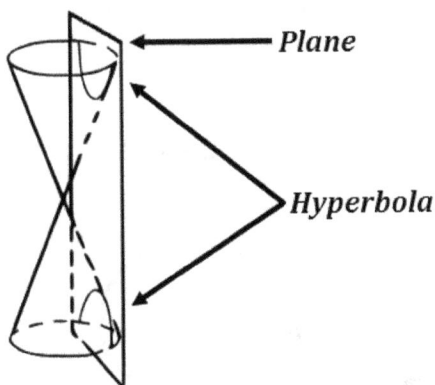

Curves

A Curve is a path traced by following a point as it moves through space. A straight line can be considered a Curve because it adheres to the aforementioned definition. A Curve can have either infinite length, such as a parabola, or a finite length depending upon the parameters set for the given Curve. If a Curve completely encloses a region of a plane, it is called a "Closed Curve". If a closed curve does not cross over itself, then it is considered to be a "Simple Closed Curve". A Circle and an Ellipse are both examples of simple closed curves.

The mathematics of Calculus examines and studies the properties of curves by examining and calculating Curves and the areas under Curves. The roots and foundation of Calculus is rest in the explanation of many ancient Geometrical problems. Some examples of Curves are presented for clarity:

Curves

Closed Curves

Simple Closed Curves

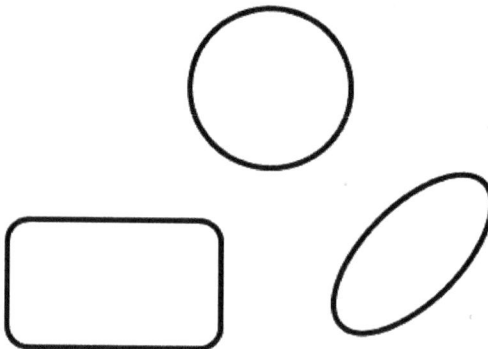

Diameter

The "Diameter" is the straight line from one side of the circumference of a circle that passes through the circle's center and meets the circumference or surface at the opposite end. The diameter effectively and equally slices the circle into two equal halves. It is illustrated as follows:

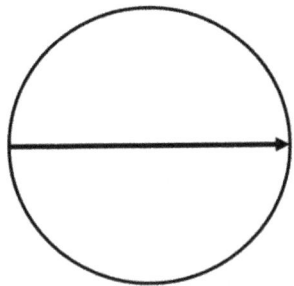

Figure

A "figure" is combination of geometric elements composed into a particular form or shape in two or three-coordinate space. Examples of two-coordinate plane figures are: the circle, the polygon, and the square. Examples of three-coordinate solid figures are: the cube, the polyhedron, and the sphere. A "Circle" is an example of a figure and is illustrated as follows:

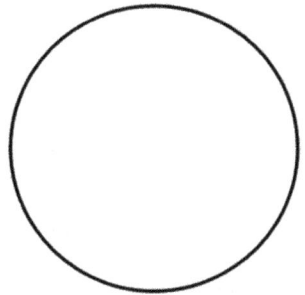

Line

A "Line" in the mathematics of Geometry can be either a sequence of points or the trace of a moving point. It is illustrated as follows:

Perimeter

The "Perimeter" of a two-coordinate figure is its border (or "outer boundary").

Pi

The lower-case Greek letter Pi: "π" (also called "pi") is a mathematical constant that applies to the circle shape. The mathematical value of pi is a ratio (or fraction) that is determined by calculating difference of a given circle's circumference [◯] divided by its diameter [◯]: $\frac{\circ}{\circ}$. The calculated value is universally applicable to any circle and is therefore considered a mathematical constant. The result of this calculation is always $\cong$ ("approximately equal to") 3.14159265... It is important to note that pi is used in many mathematical formulae in several disciplines and sciences such as: mathematics, the sciences, and engineering. Thus π, is universally considered one of the most important mathematical constants.

$$\pi = \frac{Circumference\ of\ a\ Circle}{Diameter\ of\ a\ Circle}$$

$$\pi = \frac{C}{d}$$

$$\pi = 3.14159265...$$

Where,

The Circumference (C) is the total distance around the outer edge of a circle. The Circumference of all circles is written mathematically: C = πd or C = $\pi(2r)$ or C = $2\pi r$.

The Diameter (*d*) is the length of a line segment joining two points on a circle that passes directly through the circle's center. The "Radius" (written as "r") is the distance from the center of a circle to a point on the circle's Circumference. The Diameter of all circles is equal to twice the Radius or 2r. The mathematical formula for the Diameter of a circle is d = C/π. Where, *d* = Diameter, C = circumference, and π = pi.

The following model illustrates the polar coordinates of pi starting with zero and rotating counter–clockwise around a circle:

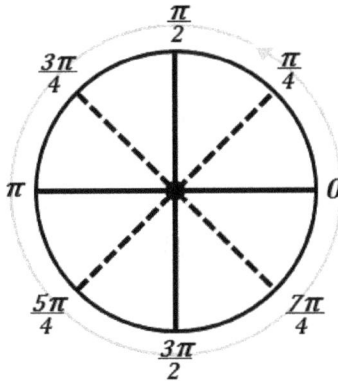

Plane

A "Plane" is a flat two-coordinate object that continues infinitely in space in all directions with no visible thickness. A plane can be defined by intersecting two lines or by creating three non–collinear points. It is illustrated using an infinitely extending parallelogram in the following manner:

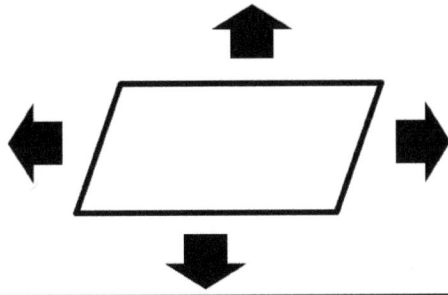

Point

The term "Point" in the mathematics of Geometry is a location that has no length, width, or depth. It is represented by a point with a dot. It is important to note the following: a dot illustrating a point has some spatial coordinates (as it was written on paper or illustrated via computer).

It is illustrated as follows:

•

Radius

The "Radius" is one half of a circle's diameter starting from the circle's center and extending to meet one side of the circle's circumference. It is illustrated as follows:

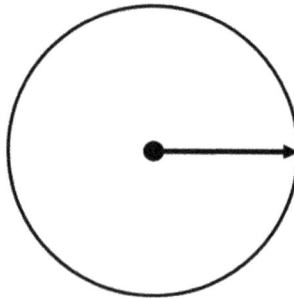

Secant

The term "Secant" in the mathematics of Geometry is an intersecting line that intersects a curve at two or more points. It is illustrated on a circle as follows:

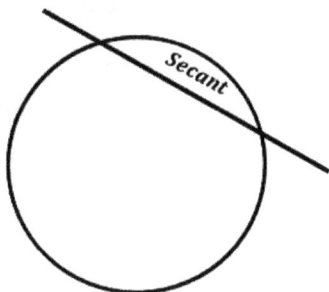

Segment

A "Segment" is a part cut off from a figure that is a finite section of a line. An arc is an example of a segment as shown in the following circle illustration between two points:

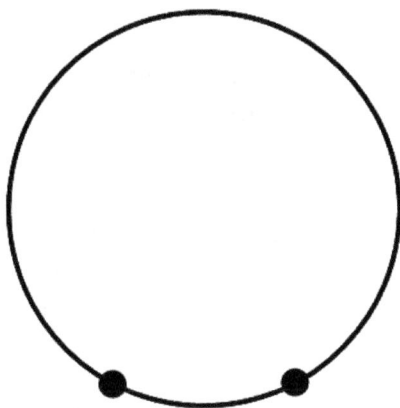

Shape

The term "Shape" in the mathematics of Geometry is a figure defined in two- or three-coordinate space by a given set of conditions. A "Circle" is an example of a shape and is illustrated as follows:

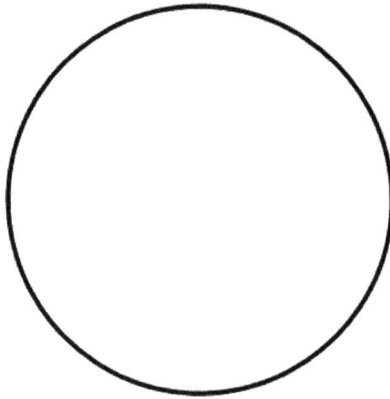

Tangent

A "Tangent" is a line or a plane that touches a curve or a surface at a single point. It is illustrated as follows:

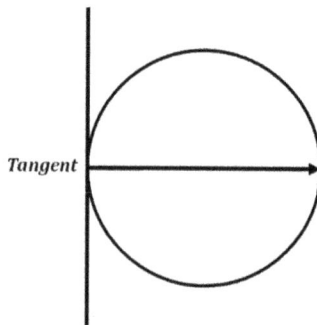

Tangent

Trigonometry the Trichotomous Opposite

The term "Trigonometry" literally means "triangle measure" it is the branch of mathematics concerned with the relations between the sides and angles of plane or spherical triangles, and the calculations based on them. Trigonometry defines these figures, their conditions, and their assumed properties in space. Trigonometry is the mathematical trichotomous opposite of the mathematical discipline of Trichometry.

The Angle Theta: ["θ"]

In mathematics the capital Greek letter Theta "θ" is most often used to represent unknown angles, especially in the study of trigonometry. Theta represents an angle in degrees, but not in radians (a radian is a unit of angular measure used in mathematics; it measures the angle of two radii as the length of the arc they create on a circle divided by the circle radius). Thus, Angle Theta ("θ") is equal to the measurement in degrees of the hypotenuse and adjacent angle in degrees (or radians) that is opposite of the 90-degree angle in right triangle.

Theta was used to represent the earth in ancient times so corresponds directly to its use in Trigonometry (the study of triangles) which is derived from Geometry (which literally means "earth measure"). The next illustration displays Angle Theta in terms of a circle in relation to the extended relationship of the angle to the circle arc and /4 radians, the Sine of Angle Theta, and the Cosine of Angle Theta:

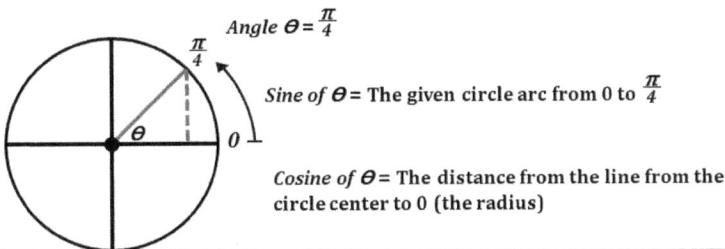

Angle $\theta = \frac{\pi}{4}$

$\frac{\pi}{4}$

Sine of θ = The given circle arc from 0 to $\frac{\pi}{4}$

Cosine of θ = The distance from the line from the circle center to 0 (the radius)

Trigonometry as the mathematics of triangles has been used to explain and illustrate elevation, the height of a structure (such as a tree or a building), and the slope of an incline (as illustrated by a right triangle which is the primary tool used in Trigonometry). Examples of how these various properties apply to right triangles are illustrated in the following 3-4-5-6 Golden Upright Right Triangle ("GURT") models:

A 1-2-3-4-5-6 or 3-4-5-6 Golden Upright Right Triangle:

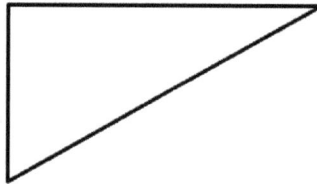

A Golden Upright Right Triangle illustrating the slope of a rapidly rising incline:

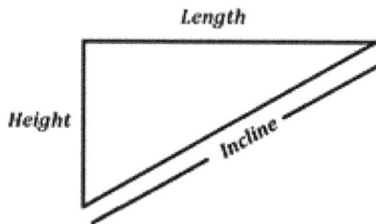

A Golden Upright Right Triangle illustrating the gradation of a slope:

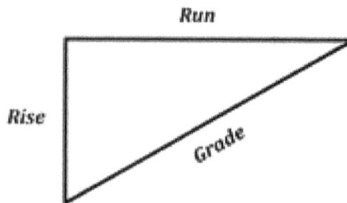

A Golden Upright Right Triangle illustrating the various sides of the shape in the terminology of Trigonometry:

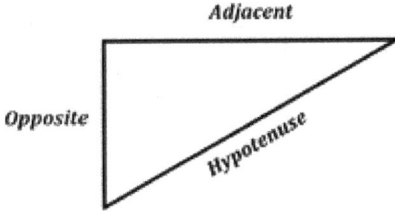

Adjacent

Opposite

Hypotenuse

A Golden Upright Right Triangle illustrating the two primary angles essential to Trigonometry:

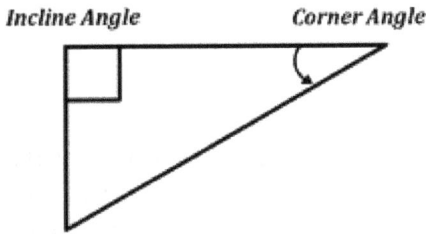

Incline Angle   **Corner Angle**

A Golden Upright Right Triangle illustrating "Angle Theta" in the terminology essential to Trigonometry:

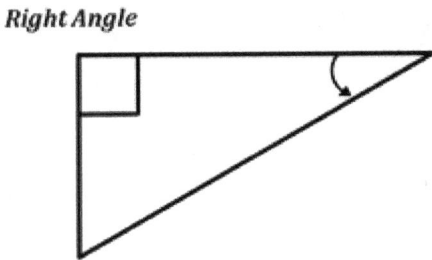

Right Angle

Angle Theta = θ

Adjacent

Opposite

Hypotenuse

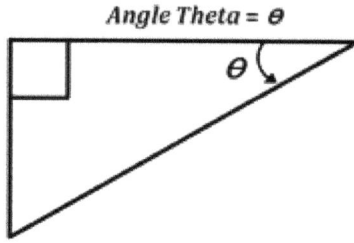

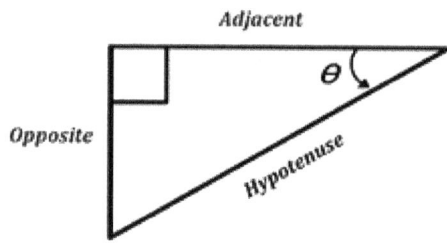

The Area around a circle also called the Circle Perimeter is called its Circumference = C. Thus, the Perimeter of a Circle = Circumference = C. The distance across a circle that runs directly through its center is called its Diameter = d.

It is important to note that the mathematical equations related to Trigonometry are constructed from the analytical creation and measurement of a right triangle from the internal structure of a circle. This is logically explained in the next series of illustrations:

Circle = = 360 Degrees = 360°.

The Golden Upright Right Triangle = = 180 Degrees = 180°.

The next series of models illustrate the how a right triangle with angle theta is constructed from a circle.

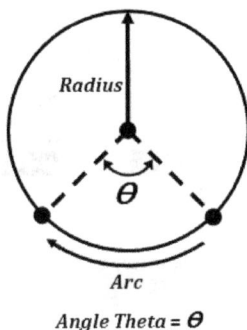

Radius

θ

Arc

Angle Theta = θ

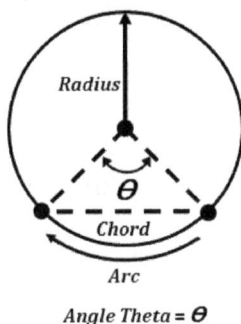

Radius

θ

Chord

Arc

Angle Theta = θ

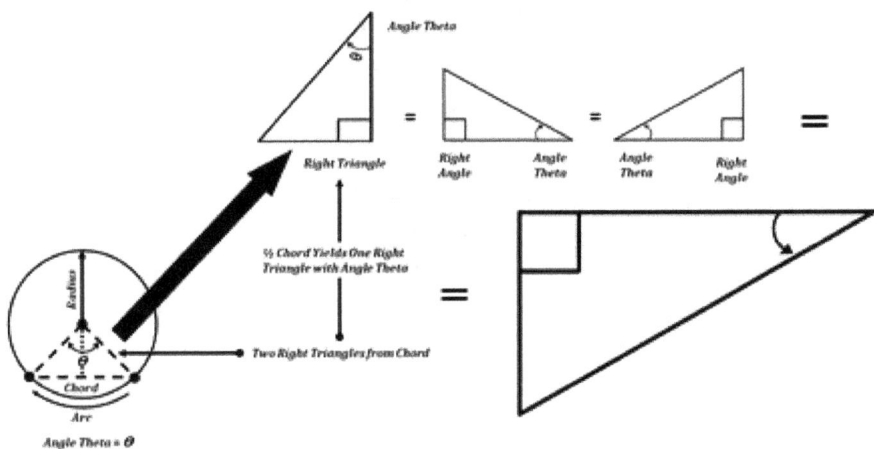

Angle Theta

Right Triangle

=

Right Angle

Angle Theta

=

Angle Theta

Right Angle

=

½ Chord Yields One Right Triangle with Angle Theta

Two Right Triangles from Chord

Radius

θ

Chord

Arc

Angle Theta = θ

=

Thus,

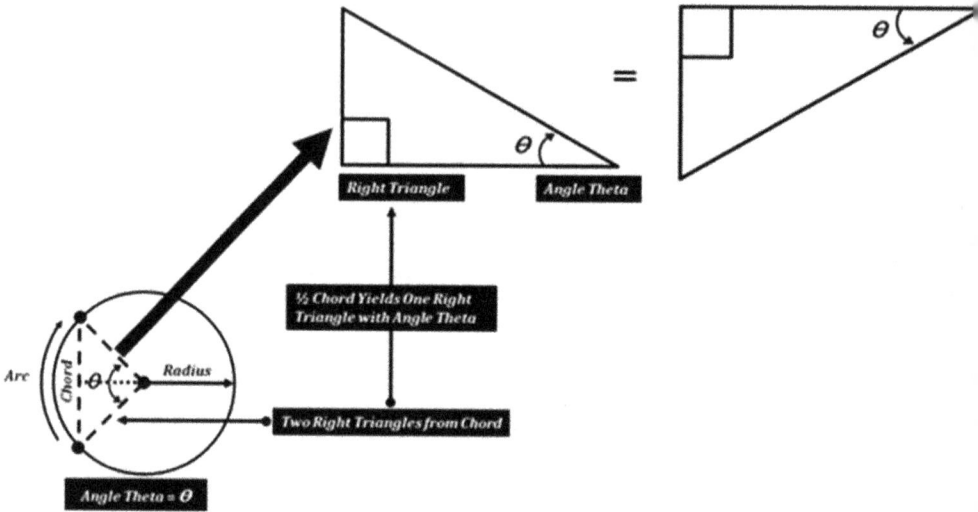

Right Triangle   Angle Theta

½ Chord Yields One Right Triangle with Angle Theta

Two Right Triangles from Chord

Arc   Chord Radius

Angle Theta = Θ

This leads to the following conclusion:

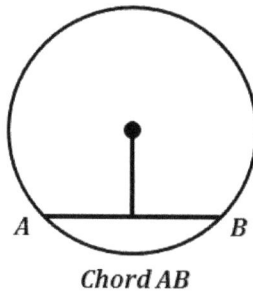

A B

Chord AB

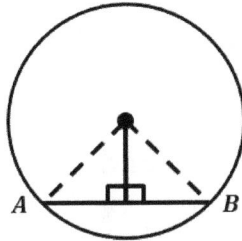

Chord AB

*Two Right Triangles
From Chord AB*

The lower-case Greek letter Pi: "π" is used to symbolize the ratio or fraction that is the Circumference over Diameter. This yields the number: **3.14159...** A universal rational number that applicable to all circles this is illustrated as follows:

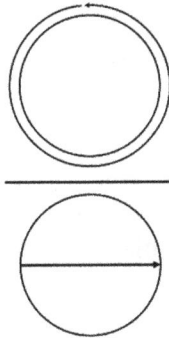

$$= \frac{C}{d} = 3.14159... = \pi$$

Where,

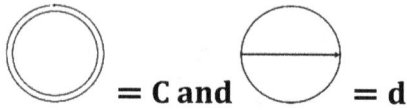 = C and  = d

Note:

 = The Circle's Perimeter

and

 = The Circle's Distance

Thus,

$$\pi = \frac{\text{Perimeter}}{\text{Distance}}$$

One half of a circle's diameter (the distance from the center of a circle to its opposite side) is called the "Radius". The radius is symbolized by the lower-case letter: "r". This is written mathematically as, **r =**

½Diameter = ½d = ½ · $\dfrac{d}{1}$ = $\dfrac{d}{2}$ ·r = . Thus, **2r = Diameter = d.**

The Area of a Circle is equal to pi times r squared: πr^2.

This is written mathematically as:

Area of a Circle = A = ⬤ = $\dfrac{\text{◎}}{\ominus} \cdot \square$ =
$\pi \cdot r^2 = \pi r^2.$

Where,

$\pi = \dfrac{\text{◎}}{\ominus}$

$r^2 = \square$, from $r = \oplus$ into $r^2 = \odot$.

Thus,

Area of a Circle = A = $\pi r^2 = $ ⬤ .

Measuring Angles

The units used to measure an angle:

Degrees

The definition of the term "Degree" in the mathematics of Trigonometry is precisely the same as it is in the mathematics of Geometry. A degree is the 360th part of a complete angle or turn (as related to a circle which has a total of 360 degrees). Each complete angle degree unit is represented by the sign "°" which immediately follows a number between 1 and 360. For example: "Thirty degrees" is written: "30°". Triangles have a total of 180 degrees when all three of the angles contained within the shape are added together.

Radians

Radians are the standard unit of measure of an angle created using the radius of a circle. Radians are used to measure angular distance in many instances rather than degrees. A single Radian unit of measurement is equivalent to the arc created by the two radii on the circumference of the circle:

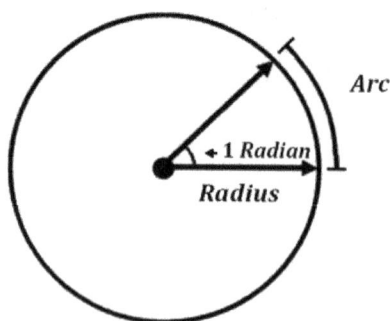

Arc Length = Radius

Trigonometric Functions in Terms of Trichometry

In mathematics, the "Trigonometric Functions" (or "Circular Functions") are the functions of an angle that is derived from a circle. These functions are used to relate the angles of a right triangle (created from the center of a given circle) to the lengths of its sides. Trigonometric Functions are important in the study of triangles, the modeling periodic phenomena, determining the height of an object from a given visual viewpoint, determining the slope of an incline, and several other applications.

The three sides of a triangle that are emphasized in Trigonometry are: the opposite side, the adjacent side, and the hypotenuse all in relation to their position relative to angle Theta. They are illustrated as follows in terms of the **Trichometry 3-4-5-6 Golden Upright Right Triangle**:

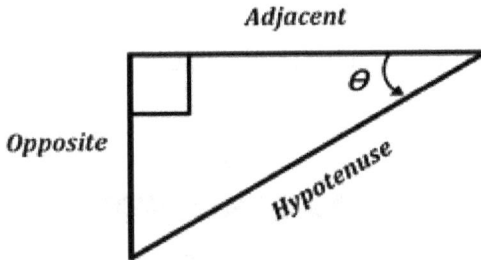

The respective position of each side of a right triangle is represented in the next series of illustrations (the sides are represented in bold):

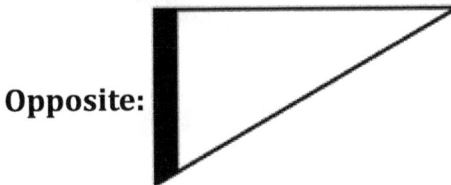

This is represented relative to angle Theta in the following manner:

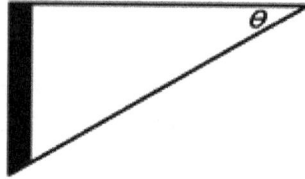

Opposite
Angle Theta = θ

Adjacent:

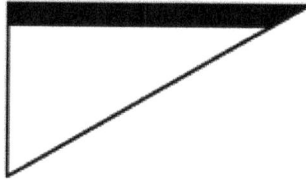

This is represented relative to angle Theta in the following manner:

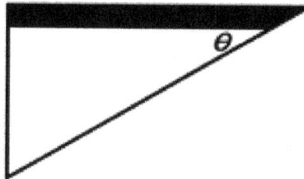

Adjacent
Angle Theta = θ

Hypotenuse:

This is represented relative to angle Theta in the following manner:

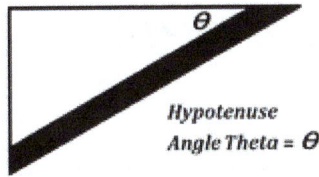

The following definitions apply to Trigonometry:

The primary three right triangle measures of angle Theta (Θ):

Sine

The "Sine" of angle Theta is the ratio of the side opposite a given Acute Angle to the Hypotenuse in a right triangle.

$$\textit{Sine of Angle } \Theta:$$

$$\textit{Sine } \Theta = \frac{\textit{Opposite}}{\textit{Hypotenuse}} = \underline{}$$

Cosine

In a right triangle the "Cosine" of angle Theta is the ratio of the side adjacent to a given angle to the hypotenuse.

Cosine of Angle Θ:

$$\text{Cosine } \theta = \frac{Adjacent}{Hypotenuse} = \underline{}$$

Tangent

In a right triangle the "Tangent" is the ratio of the side opposite a given angle to the side adjacent to the angle.

Tangent of Angle Θ:

$$\text{Tangent } \theta = \frac{Opposite}{Adjacent} = \underline{}$$

The secondary three right triangle measures of angle Theta (Θ):

Secant

In a right triangle the "Secant" is the ratio of the hypotenuse to the side adjacent to a given angle.

Secant of Angle Θ:

$$Secant\ \Theta = \frac{Hypotenuse}{Adjacent} = \frac{1}{Cosine\ \Theta} = \frac{1}{Adjacent/Hypotenuse} =$$

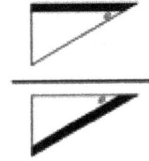

Cosecant

In a right triangle the "Cosecant" is the ratio of the hypotenuse to the side opposite a given angle.

Cosecant of Angle Θ:

$$Cosecant\ \Theta = \frac{Hypotenuse}{Opposite} = \frac{1}{Sine\ \Theta} = \frac{1}{Opposite/Hypotenuse} = 1 \div$$

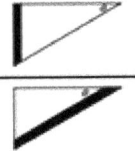

Cotangent

In a right triangle the "Cotangent" is the ratio of the side adjacent to a given angle to the side opposite.

Cotangent of Angle Θ:

$$Cotangent\ \Theta = \frac{Adjacent}{Opposite} = \frac{1}{Tangent\ \Theta} = \frac{1}{Opposite/Adjacent} = 1 \div$$

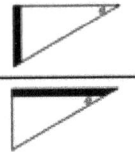

The Versine of a right triangle:

Versine

In a right triangle the "Versine" is one minus the cosine of a given angle or arc.

Examples of the aforementioned definitions are illustrated in the following model:

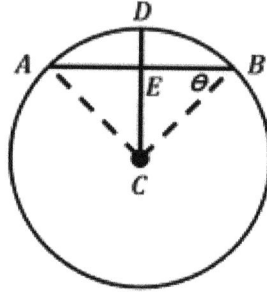

Chord = AB
Radius = CD
Hypotenuse = AC
Sine = AE
Cosine = CE
Versine = ED

Equation Symbols and Their Respective Meanings

Symbol	Mathematical Definition	Mathematical Meaning	Expression Definition	Example
( )	Parentheses	"Quantity"	Denotes A Quantity	$(x + y)$
[ ]	Square Brackets	"The Quantity"	Denotes A Quantity	$w + [(x + y) + z]$
=	Equal Sign	"Equals"	Indicates Two Values Are The Same	$-(-5) = 5$
≈	Approximate Equal Sign	"Is Approximately Equal To"	Indicates Two Values Are Close To Each Other	$x + y \approx z$
≤	Inequality Sign	"Is Not Equal To"	Indicates Two Values Are Different	$3 \leq 5$
<	Inequality Sign	"Is Less Than"	Indicates Value On Left Is Smaller Than Value On Right	$3 < 5$
≥	Inequality Sign	"Is Less Than Or Equal To" "Is At Most Equal To"	Indicates Value On Left Is Smaller Than Or Equal To Value On Right	$x \geq y$
>	Inequality Sign	"Is Greater Than"	Indicates Value On Left Is Larger Than Value On Right	$5 > 3$
≠	Inequality Sign	"Is Greater Than Or Equal To"	Indicates Value On Left Is Larger Than Or Equal To Value On Right	$x \neq y$
| |	Absolute Value Sign	"The Absolute Value of"	Distance Of Value From Origin In Number Line, Plane, Or Space	$\|-3\| = 3$

39

Equation Symbols and Their Respective Meanings

Symbol	Mathematical Definition	Mathematical Meaning	Expression Definition	Example
+	Addition Sign	"Plus"	Sum of Values	$3 + 5 = 8$
*	Multiplication Sign	"Times"	Product Of Two Values	$A * B = B * A$
x	Multiplication Sign	"Times"	Product of Two Values	$3 \times 5 = 15$
•	Multiplication Sign	"Times"	Product of Two Values	$3 \cdot 5 = 3 \times 5 = 15$
−	Subtraction Sign Minus Sign	"Minus" or "Negative"	Difference of Two Values, Negative Number	$3 - 5 = -2$
±	Plus/Minus Sign	"Plus" Or "Minus"	Expression Of Range	$500 \pm 10\%$
^	Carat	"To The Power of"	Exponent	$2\text{^}5 = 2^5 = 32$
!	Exclamation	"Factorial"	Product Of All Positive Integers Up To A Certain Value	$5! = 5 \times 4 \times 3 \times 2 \times 1 = 120$
√	Surd or Square Root Symbol	"The Root of" or "The Square Root of"	Algebraic Expressions	$\sqrt{4} = 2$
...	Continuation Sign	"And So On Up To" "And So On Indefinitely"	Extension Of Sequence	$S = \{1, 2, 3, ...\}$
/	Slash	"Divided By" "Over"	Division	$3/4 = 0.75$
÷	Division Sign	"Divided By"	Division	$3 \div 4 = 0.75$
%	Percent Symbol	"Percent"	Proportion	$0.032 = 3.2\%$
:	Ratio Sign	"Is To" "Such That" "It Is True That"	Division Or Ratio, Symbol Following Logical Quantifier Or Used In Defining A Set	$1{:}2 = 10{:}20$
∞	Lemniscate	"Infinity" "Increases Without Limit"	Infinite Summations Infinite Sequence Limit	$x < \infty$

Set Theory Symbols and Their Respective Meanings

Symbol	Mathematical Definition	Mathematical Meaning	Expression Definition	Example
( )	Parentheses	"List" "Set of Coordinates" "Open Interval"	Set Of Coordinates, or An Open Interval	$(3,5)$
[ ]	Square Brackets	"The Closed Interval" "Concentration"	Denotes A Quantity or A Closed Interval	$w + [(x + y) + z]$ $[3,5]$
( ]	Hybrid Brackets	"The Half–Open Interval"	Denotes A Half-Open Interval	$(3,5]$
[ )	Hybrid Brackets	"The Half-Open Interval"	Denotes A Half-Open Interval	$[3,5)$
{ }	Curly Brackets	"The Quantity" "The Set"	Denotes A Quantity Or A Set	$E = \{2, 4, 6, 8, ...\}$
|	Vertical Line	"Such That" "It Is True That"	Symbol Following Logical Quantifier Or Used In Defining A Set	$S = \{x \mid x < 2\}$
$\in$	Element–Of Symbol	"Is An Element Of A Set"	Sets	$a \in A$
$\notin$	Not–Element–Of Symbol	"Is Not An Element Of A Set"	Sets	$b \notin A$
$\subseteq$	Subset Symbol	"Is A Subset Of"	Sets	$A \subseteq B$
$\subset$	Proper Subset Symbol	"Is A Proper Subset Of"	Sets	$A \subset B$
$\cup$	Symbol	""	Sets	$A \cup B = B \cup A$
$\cap$	Intersection Symbol	"Intersect"	Sets	$A \cap B = B \cap A$
$\varnothing$	Null Symbol	"The Null Set" "The Empty Set"	Sets	$\varnothing = \{\}$

Chapter Two follows and presents Trichometry Terminology.

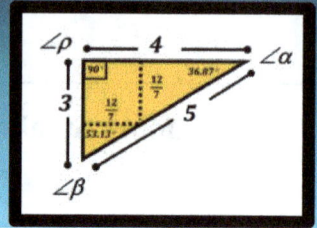

Every good gift and every perfect gift is from above, and cometh down from the Father of lights, with whom is no variableness, neither shadow of turning.

James 1: 17

<u>A Glossary of Terminology</u>
<u>Relevant to Trichometry</u>

Algorithm

An Algorithm is a sequence of finite events or a sequence of finite instructions. In Trimensional Analysis, Perceptology, and Visualus an "Algorithm" is a sequence of finite events or a sequence of finite instructions. Algorithms are constructed as a concise means of explaining Perceptological concepts. This is of great value when describing the events that take place when one learns interactively.

In Visualus an "Algorithm" is a sequence of finite events or a sequence of finite instructions. Algorithms are constructed as a means of explaining Visualus Solutions. The "Visualus Front Face" of the "Visualus Isometric Cuboid" that is the "Analysis Rectangle" and is the 3 by 3 Standard Table of the Triostatistic Tri–Squared Test that is halved to create the "Trichometry 3-4-5-6 Golden Upright Right Triangle".

Analysis Rectangle

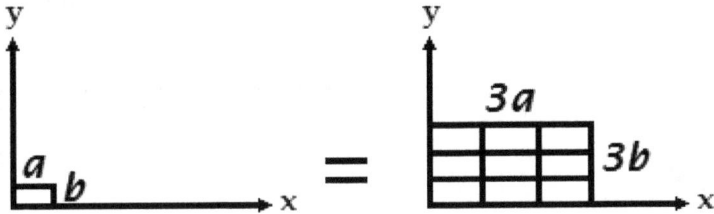

Note the following:

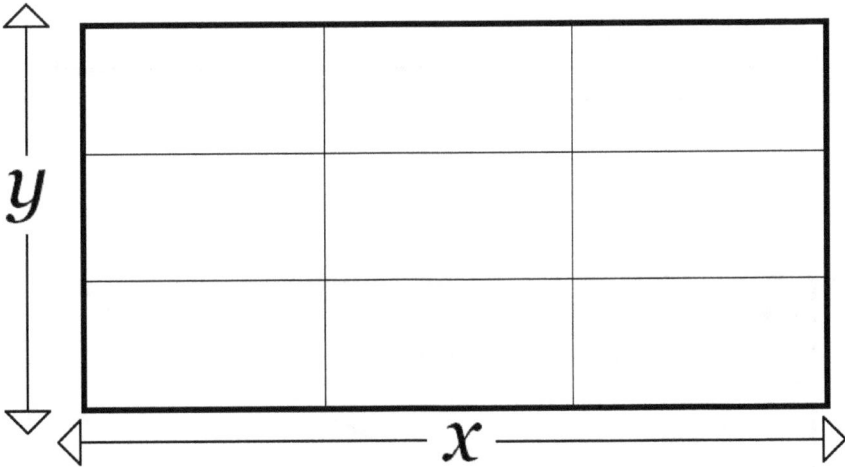

The Analysis Rectangle is based upon a two-coordinate Cartesian Coordinate rectangle ($a \cdot b = ab$) that is dilated by a factor of 3 into the first Visualus 9 Categorical Area Analyze Vector of the Innovative Problem–Solving Model of Inventive Instructional Design (mathematically represented as: $3a \cdot 3b = 3a3b = 9ab$). The transition into the 3-4-5-6 GURT is more clearly illustrated in the next series of models.

The creation of the "Trichometry 3-4-5-6 Golden Upright Right Triangle" from the "3 by 3 Visualus Isometric Cuboid Front Face" is more clearly illustrated in the next two Cartesian Coordinate graphical models.

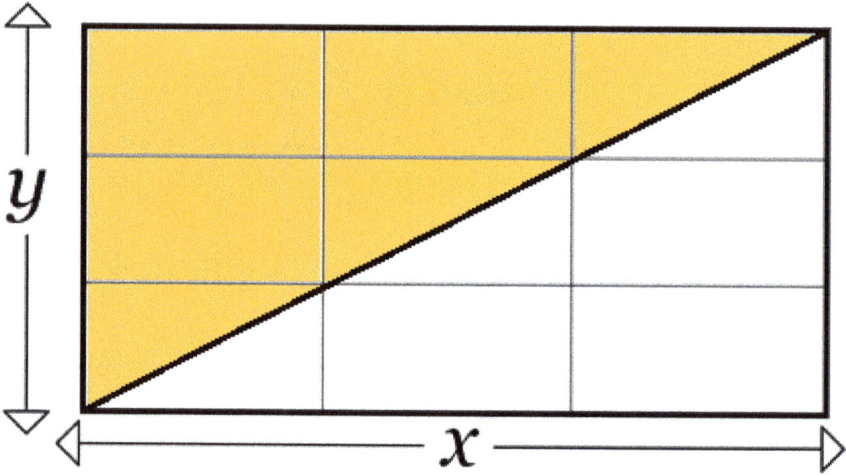

Into

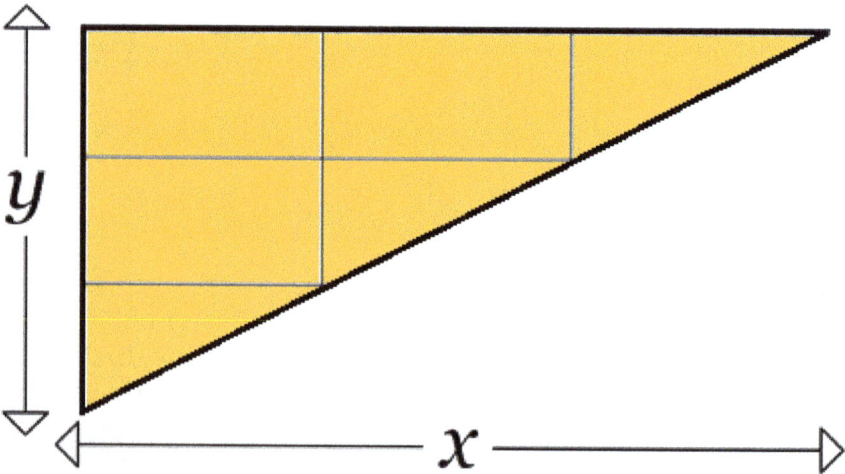

The "∇" [The "Trichotomous Upright Right Triangle"] as the Foundation of All of the Triological Sciences

The Triological Sciences (which are identified as—(a.) Triology; (b.) Trithmetic; (c.) Triomathematics; (d.) Triophysics; (e.) Triostatistics; (f.) Trioengineering; (g.) Trichometry; and (h.) Triangulus) are all grounded by the same foundation that has its roots in the geometric "Golden Upright Right Triangle" also known as the "3-4-5-6 Golden Upright Right Triangle" (represented by the acronym "GURT"). The Golden Upright Right Triangle is derived as an internal characteristic from the "Visualus Isometric Cuboid © ™" and is symbolized by the "Trine abc" = "∇abc". The Trine symbol is also the mathematical symbol for the "Trichotomous Upright Right Triangle" as the triple terms "Trichotomous Upright Right Triangle" as [∇ = "The Trichotomous Upright Right Triangle" or "Upright Right Trine"]. Therefore, the appearance of the Trine symbol before a title in the Triological Sciences indicates a shorthand notation for the words "Trichotomous Upright Right Triangle" directly referring to the "Golden Upright Right Triangle" (as such "Trine *abc*" = "∇abc" = "Golden Upright Right Triangle of *abc*"). The true foundation of all of the "Triological Sciences" is the base underpinning of mathematical model of Visualus which is the Isometric Cuboid. The Visualus Isometric Cuboid is presented graphically below:

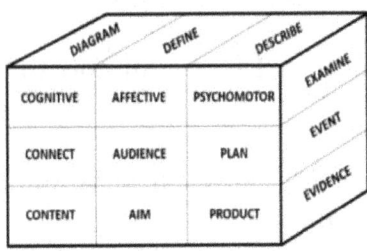

"Visualus" is the mathematics of innovative, inventive, and in-depth problem-solving via the geometric representation of the "Rectilinear Model of Instructional Systems Design". The next series of models and equations with explain in explicit detail and meticulous precision exactly how *V̷abc* is derived from the Visualus Isometric Cuboid as the inherent foundation and true beginning of all of the Triological Sciences. The Trichotomous Upright Right Triangle of *abc* is defined in Cartesian Coordinates mathematically in the following "Golden Upright Right Triangle Triological Side to Coordinate Mathematical Definition Equation":

$$V̷abc = V̷[abc] = V̷[xyz] = V̷[x] \cdot V̷[y] \cdot V̷[z] = V̷[a] \cdot V̷[b] \cdot V̷[c] \text{ for } V̷[\text{Side } a] \cdot V̷[\text{Side } b] \cdot V̷[\text{Side } c] = V̷abc.$$

According to the above " V̷Triological Side to Coordinate Mathematical Definition Equation", the following applies:

The holistic all-inclusive "GURT" is represented by " V̷ " = *Trioengineering* = Trine;
The measurable unit "Length" is represented by "*x*" = *x*–coordinate = abscissa;
The measurable unit "Height" is represented by "*y*" = *y*–coordinate = ordinate;
The measurable unit "Depth" is represented by "*z*" = *z*–coordinate = applicate; and
The mathematical operator "[ ]" is equitable to "a concentration on" = "to focus on".

Graphically, the relationship between *V̷abc* and the Visualus Isometric Cuboid is illustrated as follows:

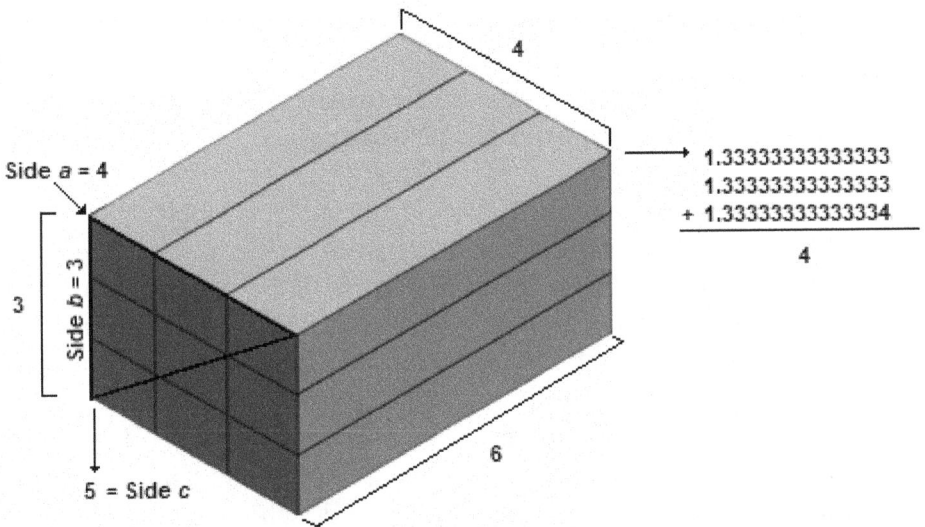

Side a = 4

4

1.33333333333333
1.33333333333333
+ 1.33333333333334
4

3

Side b = 3

5 = Side c

6

Thus, from $\sqrt{abc}$ which is extracted out of the Visualus Isometric Cuboid: $[xyz] = [x][y][z]$ illustrating the three Cartesian Coordinates. Idealistically, in terms of vectors, if Cartesian Coordinates x and z are respectively perpendicular then their cross product is Cartesian Coordinate y. The reduction of Cartesian Coordinate z to present only the Front Face of the Visualus Isometric Cuboid that is also the 3 by 3 Standard Table Format of the Triostatistics Tri–Squared Test. The mathematics of the removal of Cartesian Coordinate z to have only the Front Face of the Isometric Cuboid as the 3 by 3 Table remain is illustrated mathematically and graphically in the sections that follow.

Defining Triological Science Notation: "Total Trioengineering Notation" or "Trioengineering Notation"

"Trioengineering Notation" is used to mathematically parsimoniously explain the "mathematical law of trichotomy" through the use of 3-4-5-6 Golden Upright Right Triangle (or "GURT") and its properties as a holistic mathematical operation in the same manner as Summation or "Sigma" Notation (that uses "Σ") and Product Notation (that uses "Π"). Trioengineering Notation has a variety of uses and applications in the various Tiological Sciences. The symbol for all Trioengineering Notation is the "Triune" or " ▽ " (that when used in Trioengineering Notation literally means "Trioengineered" which is a shorthand way of saying "The Trioengingeering 3-4-5-6 Golden Upright Right Triangle"). Whenever and wherever the Trioengineered Trioengineering Notation Triune " ▽ " symbol is used it indicates that the equation, model, solution, and/or calculation uses and is a part of Trioengineering as a whole via the 3-4-5-6 Golden Upright Right Triangle. As the GURT is the universal and connective foundation of all of the Triological Sciences and all of Trichotomous Research—Trioengineering Notation is therefore universally used).The notation has broad utility and used in multiple ways using the following formats that are detailed in all of the sections and equations, models, solutions, and calculations that follows.

Trioengineering Notation

Trioengineering Notation has the same nomenclature parameters as the more traditional "Summation Notation" that uses the upper case Greek letter "Sigma" ("Σ") to indicate summation in a series and "Product Notation" that uses the upper case Greek letter "Pi" ("Π") to indicate multiplication in a series. Trioengineering Notation is used in this format connote geometric trichotomous equations, calculations, and formulae that directly pertain to the utility and viability of the 3-4-5-6 Golden Upright Right Triangle in problem-solving and the Triological Sciences in particular. In this format, Trioengineering Notation is written as follows to connote the triple trichotomy of a concept, idea, thought and/or solution (that corresponds directly to the triune trichotomy of the three angles and three sides of the GURT) as follows:

$$\overset{\displaystyle 3}{\underset{\displaystyle i=1}{\triangledown}}$$

Where,

$\triangledown$ = The "**Triune**" Symbol that literally means "**Trioengineered**" or "**Trioengineering**" = "The Trichotomous Upright Right Triangle" = "The 3-4-5-6 Golden Upright Right Triangle" (or "GURT");
3 = The immutable unchanging ending point called a "**Triand**" for "Trichotomous Combination End"; and
1 = The starting point called a "**Triart**" for "Trichotomous Start" (that always at $\angle\alpha$ and/or Side a depending upon the model, for example Triostatistics Triangular Equation Modeling or [TEM] that always starts with "Side a" versus the cyclical Trioengineering Model of ["author"; "build"; and "convey"] that always starts with "$\angle\alpha$") on the $\triangledown$ even when the model is cyclical and ongoing; and
i = The index (or the "Starting Point") that is always at 1 on the $\triangledown$.

Trioengineering Notation is used in geometrical names to connote the actual application of the "3-4-5-6 Golden Upright Right Triangle" in its entirety, as in this example:

" ∇abc" (that can be literally defined and referred to as "Triune abc" or "Trioengineered abc").

It is also used in mathematical formulas and equations as illustrated in this example:

$$\nabla[\; \nabla y = \nabla mx + \nabla b].$$

Used as the preconditional modifier that explains the application of the Golden Upright Right Triangle in defining in-depth and precise mathematical operations as GURT conditions (using both or either "Tri" and " ∇") as exhibited below:

$$\underset{name}{\overset{n=3}{\mathrm{Tri}}} = \underset{name}{\overset{Tri}{\nabla}} ; and$$

$$\underset{name}{\overset{n=3}{\mathrm{Tri}}}[calculation\ and/or\ formula] = \underset{name}{\overset{Tri}{\nabla}}[calculation\ and/or\ formula].$$

Additionally, Trioengineering Notation is used to convey the Tripositive nature of all positive integers as defined in explicit detail in the next section. Trioengineering format when used to define all positive numbers (including zero) in this manner is written and expressed as follows:

$$\underset{i=1}{\overset{n=3}{\underset{num}{\nabla}}} .$$

The Final Full and Complete Mathematical Definition of Trioengineering Notation

Trioengineering Notation is fully and completely mathematically defined in the following definitive mathematical series of equations presented below.

Trioengineering is meticulously, precisely, specifically used to parsimoniously define the trichotomous use of the 3-4-5-6 Upright Right Triangle in the following mathematical equations:

$\nabla = \overset{3}{\underset{i=1}{\nabla}}$ = Vertical Abbreviation ∇ (with notation on the Trine) = The definition of all Positive Integers (including Zero because it completely satisfies the definition and conditions of Upright Right Triangle numbers) Upright Triangle Number applicable to the measurements = ∇ Name or Abbreviation of Measure = [Tri] as the "Tripositive Trichotomy" where, [∇ = "Tri"] = The Geometric Models as the ""CTCG Function"" = where CTCG = ["Convert" to by "Transforming" into and "Conforming" to "Gain"] = The "3-4-5-6 Golden Upright Right Triangle" that is also referred to as the "Trichotomous Upright Right Triangle" geometrically to =

= by CTCG into = $\nabla abc \equiv$ (m = 1 for a measurement of 3 = 4/3 cells on the x-axis for the 3 by 3 Table) into

(m = 0.75 for a measurement of 4 = 3/3 cells on the x-axis

for the 3 by 3 Table) at $\left.\begin{array}{l} a = 4; \\ b = 3; \\ c = 5; \text{ and} \\ A = 6. \end{array}\right\}$

Triological Science Triomathematics: Trichotomous Upright Right Triangle Numbers

Johann Carl Friedrich Gauss was a noted and well known 18th Century German mathematician and physicist who made significant contributions to many fields in both math and science. He is universally considered to be one of history's most influential mathematicians. In Gauss's diary entry on July 10th, 1796 he made a great discovery related to sum of triangular numbers that stated: "**EYPHKA! num = Δ + Δ + Δ**" and in addition records his discovery of a proof that any number is the sum of 3 triangular numbers. This is more commonly known as Gauss's Eureka Theorem. The Theorem is modified by the author into the "Trichotomous Upright Right Triangle" Theorem and uses Trioengineering Notation to rewrite the original theorem in the context of the Trichotomous Upright Right Triangle into the following:

$$\sum_{i=1}^{n=3} \triangledown = \triangledown_1 + \triangledown_2 + \triangledown_3 \ .$$

The above "Trioengineering Notation Trichotomous Upright Right Triangle Equation" defines all positive numbers as true "Tripositive Trichotomous Upright Right Triangle Integers" (including the number "0"). As such, all "Tripositive Trichotomous Upright Right Triangle Integers" have the following trichotomous properties: (1.) "Magnitude" (as "Size"); (2.) "Distance" (as "Length"); and (3.) "Position" (as "Place") in a self-contained inharmonic summative "***Triharmonic Triune***". These numeric properties are valid characteristics that are reflected in nature as "Triologically Scientifically" viable and tangible with an intentional sub-atomic infrastructure that is the initial structural framework for all of existence, all of reality, and all of nature.

All positive integers including zero "0" are "Trichotomous Upright Right Triangle Numbers" and their respective values can all be trichotomously created in a trifold manner via a threefold Tripositive calculation. The Tripositive calculation is satisfied by the equation: $[(n^2 + n)/2]$, which is the mathematical formula used to create the Triangular Numbers need to complete all positive integers including zero as "Trichotomous Upright Right Triangle Numbers". The Tripositive is... The Tripositive is written in the following manner: "[value]". In the case of the "Trichotomous Upright Right Triangle Numbers", the Tripositive number is written as "[num]" and is parsimoniously rewritten as, "[n]", thus, "[num] = [n]", for all positive integers including zero ("0").

The Golden Upright Right Triangle Vector Model displaying how the 3-4-5-6 Golden Upright Right Triangle positing conveys the Pythagorean Theorem to Produce the Acclivity of Side c:

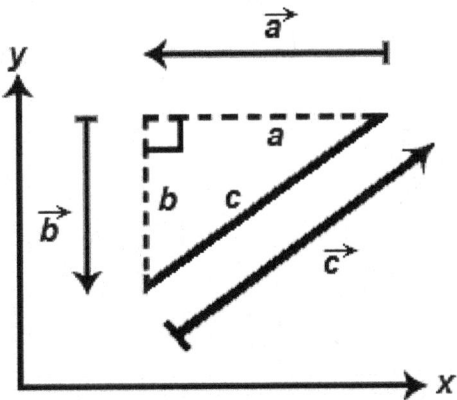

Where,
$a = x$
$b = y$

For all Unit Values as Vectors that become the 3-4-5-6 GURT Sides respectively.

Note:

If $a = \begin{bmatrix} x \\ y \end{bmatrix}$ then the size of $|c| = \sqrt{(x^2 + y^2)}$

Side c can be mathematically expressed as a unit value in the 3-4-5-6 Golden Upright Right Triangle as the "Total Side c [3-4-5-6 GURT] Equation" as a mathematical formula:

$$Side\ c = Side\ a\left[\sqrt{1 + \left(\frac{Side\ b}{Side\ a}\right)^2}\right].$$

The aforementioned can be further simplified into a more "Complete Form" of the "Total Side c Equation" that is written in the following manner:

$$c = a\left[\sqrt{1 + \left(\frac{b}{a}\right)^2}\right].$$

Upon the input of the three respective 3-4-5-6 GURT Side Values as Side $a = 4$; Side $b = 3$; the final "Calculated Form" of the "Total Side c Equation" will yield Side $c = 5$ as illustrated below:

$$5 = 4\left[\sqrt{1 + \left(\frac{3}{4}\right)^2}\right].$$

According to the abovementioned 3-4-5-6 GURT Vector Model, "the acclivity" (which literally means "the upward slope" of Side c = Side b over Side a) of side c is mathematically defined as $b/a = ¾ = 0.75$. In the 3-4-5-6 Golden Upright Right Triangle Side $a = 4$; Side $b = 3$; and Side $c = 5$ (therefore as illustrated in the model above the respective x and y coordinate abscissa and ordinate as vectors {a line with both direction and magnitude} measures out to x = Side a = 4 Unit Side Vector = $\vec{a}$, with y = Side b = 3 as the Unit Side Vector = $\vec{b}$, and [x, y] = Side c = 5 as the Unit Side Vector = $\vec{c}$, respectively, therefore, the equation, $|c| = \sqrt{(x^2 + y^2)}$ is satisfied because, $|5| = \sqrt{(4^2 + 3^2)} = \sqrt{(16 + 9)} = \sqrt{25} = 5$); with an overall area of A = 6. The 3-4-5-6 Golden Upright

Right Triangle is the basis for the Triological Science Triostatistics "Triangular Modeling Equation also known as "[TEM]". [TEM] is used to model a vast variety of trichotomous relations and relationships between ideas, concepts, and solutions in research. The 3-4-5-6 Golden Upright Right Triangle characteristics are as follows: .

Triangle Numbers Expressed via the Total Triangle Notation that is "Trioengineering Notation"

The Golden Upright Right Triangle Identity for all Positive Integers (including Zero) written in "Upright Right Total Triangle Notation" in following manner:

$$\sum_{i=1}^{n=3} \triangledown \equiv \triangledown_1 + \triangledown_2 + \triangledown_3 \,.$$

The solution above is based on mathematician Carl Gauss' solution from his 1796 notebook which stated the following as the first presentation of the Triangle Number Theorem as a Triangular Number proof which originally was entered on 07.10.1796 and read as follows (note: "EYPHKA!" written in Greek which is translated into English as: "Eureka!") was originally written as follows:

EYPHKA! num = $\triangledown$ + $\triangledown$ + $\triangledown$,

The "Mathematical Identity for Upright Right Triangle Numbers" (which provides each Upright Right Triangle Number with a mathematical definition that is equivalent to Gauss' original Triangle Number Proof. An "Upright Right Triangle Number" (for all positive integers is represented by the following notation and symbol: "$\sum_{i=1}^{n=3} \triangledown$ ".

The aforementioned is the notation for all positive integers (including zero) that are composed of the sum of three Upright Right Triangle Numbers. The symbol " ∇ " (an italicized "nabla") is used in this case to indicate the "Golden Upright Right Triangle" (or "GURT") as was first introduced by the author in the 2021 referred journal article entitled, "Tri–Power Analysis: The Advanced Post Hoc Triostatistical Assurance Model that Consists of Multiple Advanced Triostatistics to further Verify, Validate, and Make Viable Ideally Replicated Results of Innovative Investigative Inquiry" in the Journal of Creative Education. It is important to note that the Upright Right Triangle Notation is equitable to the following equation which denotes its intrinsic equality to Gauss' original Eureka Theorem: [$\nabla + \nabla + \nabla = \triangle + \triangle + \triangle =$ num]; within the confines of the upright right triangle identity "[n² + n] ÷ 2" (is a calculation which equates directly to the traditional equation "Triangle Number Theorem" as: $\frac{n(n+1)}{2}$) is now written as the dual identity:

$$\sum_{i=1}^{n=3} \nabla \equiv \text{Tri[n]}_{\nabla} \equiv [n^2 + n] \div 2$$

as

$$\sum_{i=1}^{n=3} \nabla = \text{Tri[n]}_{\nabla} = n[n + 1] \div 2 = [\text{num}] = [n];$$ where, "n(n + 1)$\big/$2" = "Triangle Number Equation"; and

$$"\sum_{i=1}^{n=3} \nabla = \text{Tri[n]}_{\nabla} = n[n + 1] \div 2 = [\text{num}] = [n]"$$ is the Upright Right "Triangle Number Theorem".

Table 1 follows and illustrates integers 1 through 5 as "Upright Right Triangle Numbers".

Table 1
The Upright Right Triangle Positive Integer Table Exhibiting Values 1 to 5

The Tabular Proof of the above as an adaptation of Knott's 2003 Triangle Number Table that Presents the Triangle Numbers in the Golden Upright Right Triangle Model Form

[n] = number	1	2	3	4	5	...
Tri[n]$_\nabla$ as a sum	1	1+2	1+2+3	1+2+3+4	1+2+3+4+5	...
All numbers 2 and higher expressed as Tri[n] as Upright Right Triangles or Tri[n]$_\nabla$	•	•• •	••• •• •	•••• ••• •• •	••••• •••• ••• •• •	...
Final Answer to: Tri[n]$_\nabla \equiv [n^2 + n] \div 2$	1	3	6	10	15	...

All numbers or integers have a base trichotomy, for example the number 1 has: +1; -1; and non-1 or zero. In addition, all numbers or integers are triangular in their separate trichtomies as illustrated in the Table above (based upon the original work of Knott, 2003).

The Upright Right Triangle Positive Integer Grid Exhibiting Values 1 to 28: The Upright Right Triangle Number Grid—Adding Positive Integers According to Base Column Number Across Rows that Creates the Shape of the 3-4-5-6 GURT

↓28	↓21	↓15	↓10	↓6	↓3	←+1
↑27	↑20	↑14	↑9	↑5	←↑+2	—
↑25	↑18	↑12	↑7	←↑+3	—	—
↑22	↑15	↑9	←↑+4	—	—	—
↑18	↑11	←↑+5	—	—	—	—
↑13	←↑+6	—	—	—	—	—
↑+7	—	—	—	—	—	—

Note that the shape of the numbers above fall into the overall form of the Golden Upright Right Triangle.

All Positive Integers including Zero as Upright Right Triangle Numbers

All positive integers including zero "0" are "Upright Right Triangle Numbers" and their respective values can all be trichotomously created in a trifold manner via a threefold Tripositive calculation. The Tripositive calculation is satisfied by the equation: $[(n^2 + n)/2]$, which is the mathematical formula used to create the Triangular Numbers need to complete all positive integers including zero as "Upright Right Triangle Numbers". The Tripositive is... The Tripositive is written in the following manner: "[value]". In the case of the "Upright Right Triangle Numbers", the Tripositive number is written as "[num]" and is parsimoniously rewritten as, "[n]", thus, "[num] = [n]", for all positive integers including zero ("0").

In deference to the aforementioned, the following is then true for the "Positive Triangular Number Values" that are used to determine all numbers as "Upright Right Triangle Numbers" are indicated as follows using the "Upright Right Triangle Number Equation" of "$[(n^2 + n)/2]$" (using zero "0" and the positive integers 1 through 9 in the "Upright Right Triangle Number Equation") are presented in the Table below.

The Positive Upright Right Triangle Integer Table Exhibiting Values 0 to 28

"Positive Right Triangle Values" (including Zero) for the Triangular Numbers Equations	The " Upright Right Triangle Number Equations" for the "Positive Right Triangle Values"	Final Numerical Outcomes
For, n = 0	Then, 0 = $[(n^2 + n)/2]$ = $[(0^2 + 0)/2]$	0/2 = 0
For, n = 1	Then, 1 = $[(n^2 + n)/2]$ = $[(1^2 + 1)/2]$	2/2 = 1
For, n = 3	Then, 3 = $[(n^2 + n)/2]$ = $[(2^2 + 2)/2]$	6/2 = 3
For, n = 6	Then, 6 = $[(n^2 + n)/2]$ = $[(3^2 + 3)/2]$	12/2 = 6
For, n = 10	Then, 10 = $[(n^2 + n)/2]$ = $[(4^2 + 4)/2]$	20/2 = 10
For, n = 15	Then, 15 = $[(n^2 + n)/2]$ = $[(5^2 + 5)/2]$	30/2 = 15
For, n = 21	Then, 21 = $[(n^2 + n)/2]$ = $[(6^2 + 6)/2]$	42/2 = 21
For, n = 28	Then, 28 = $[(n^2 + n)/2]$ = $[(7^2 + 7)/2]$	56/2 = 28
For, n = 36	Then, 36 = $[(n^2 + n)/2]$ = $[(8^2 + 8)/2]$	72/2 = 36
For, n = 45	Then, 45 = $[(n^2 + n)/2]$ = $[(9^2 + 9)/2]$	90/2 = 45
	...and the pattern continues...	

The abovementioned Triangular Numbers as the "Final Numerical Outcomes" in the Table can then be used to repetitively create all positive numbers in groups of 3 to illustrate that all positive numbers (positive integers) are Triangular Numbers. Table 4 follows and illustrates **"The Upright Right Triangle Positive Integer Table Exhibiting Values 0 to 45":**

"Positive Integer" (including Zero)	The "Positive Right Triangle Value Equations" Using the "Positive Right Triangle Values"	Final Positive Integer
For, n = 0	[0] + [0] + [0] = 0	0
For, n = 1	[0] + [0] + [1] = 1	1
For, n = 2	[0] + [1] + [1] = 2	2
For, n = 3	[1] + [1] + [1] = 3	3
For, n = 4	[0] + [1] + [3] = 4	4
For, n = 5	[1] + [1] + [3] = 5	5
For, n = 6	[0] + [3] + [3] = 6	6
For, n = 7	[1] + [3] + [3] = 7	7
For, n = 8	[1] + [1] + [6] = 8	8
For, n = 9	[0] + [3] + [6] = 9	9
For, n = 10	[0] + [0] + [10] = 10	10
For, n = 11	[0] + [1] + [10] = 11	11
For, n = 12	[1] + [1] + [10] = 12	12
For, n = 13	[0] + [3] + [10] = 13	13
For, n = 14	[1] + [3] + [10] = 14	14
For, n = 15	[0] + [0] + [15] = 15	15
For, n = 16	[0] + [1] + [15] = 16	16
For, n = 17	[1] + [1] + [15] = 17	17
For, n = 18	[0] + [3] + [15] = 18	18
For, n = 19	[1] + [3] + [15] = 19	19
For, n = 20	[0] + [10] + [10] = 20	20
For, n = 21	[0] + [0] + [21] = 21	21
For, n = 22	[0] + [1] + [21] = 22	22
For, n = 23	[1] + [1] + [21] = 23	23
For, n = 24	[0] + [3] + [21] = 24	24
For, n = 25	[1] + [3] + [21] = 25	25
For, n = 26	[1] + [10] + [15] = 26	26
For, n = 27	[6] + [6] + [15] = 27	27
For, n = 28	[0] + [0] + [28] = 28	28
For, n = 29	[0] + [1] + [28] = 29	29
For, n = 30	[0] + [15] + [15] = 30	30
For, n = 31	[0] + [3] + [28] = 31	31
For, n = 32	[1] + [3] + [28] = 32	32
For, n = 33	[6] + [6] + [21] = 33	33
For, n = 34	[0] + [6] + [28] = 34	34
For, n = 35	[1] + [6] + [28] = 35	35
For, n = 36	[0] + [0] + [36] = 36	36
For, n = 37	[0] + [1] + [36] = 37	37
For, n = 38	[1] + [1] + [36] = 38	38
For, n = 39	[0] + [3] + [36] = 39	39
For, n = 40	[6] + [6] + [28] = 40	40
For, n = 41	[3] + [10] + [28] = 41	41
For, n = 42	[0] + [6] + [36] = 42	42
For, n = 43	[0] + [15] + [28] = 43	43
For, n = 44	[6] + [10] + [28] = 44	44
For, n = 45	[0] + [0] + [45] = 45	45

...and the pattern continues...

In summary and deference to the aforementioned, the following is then true for the "Positive Triangular Number Values" that are used to determine all numbers as "Trichotomous Upright Right Triangle Numbers" are indicated as follows using the "Trichotomous Upright Right Triangle Number Equation" of "$[(n^2 + n)/2]$" (using zero "0" and the positive integers 1 through 9 in the "Trichotomous Upright Right Triangle Number Equation") written in a "non-tabular formulaic format" as follows:

For n = 0, then 0 = $[(n^2 + n)/2]$ = $[(0^2 + 0)/2]$ = 0/2 = 0;
For n = 1, then 1 = $[(n^2 + n)/2]$ = $[(1^2 + 1)/2]$ = 2/2 = 1;
For n = 3, then 3 = $[(n^2 + n)/2]$ = $[(2^2 + 2)/2]$ = 6/2 = 3;
For n = 6, then 6 = $[(n^2 + n)/2]$ = $[(3^2 + 3)/2]$ = 12/2 = 6;
For n = 10, then 10 = $[(n^2 + n)/2]$ = $[(4^2 + 4)/2]$ = 20/2 = 10;
For n = 15, then 15 = $[(n^2 + n)/2]$ = $[(5^2 + 5)/2]$ = 30/2 = 15;
For n = 21, then 21 = $[(n^2 + n)/2]$ = $[(6^2 + 6)/2]$ = 42/2 = 21;
For n = 28, then 28 = $[(n^2 + n)/2]$ = $[(7^2 + 7)/2]$ = 56/2 = 28;
For n = 36, then 36 = $[(n^2 + n)/2]$ = $[(8^2 + 8)/2]$ = 72/2 = 36;
For n = 45, then 45 = $[(n^2 + n)/2]$ = $[(9^2 + 9)/2]$ = 90/2 = 45;

The next series of Triangular Numbers is 10, 15, and 21 respectively to repeat the pattern and continue to define all positive integers as Triangular Numbers. Written in a "non-tabular full format" as follows:

Where, n = 10, then 10 = $[(n^2 + n)/2]$ = $[(4^2 + 4)/2]$ = 20/2 = 10; and n = 15, then 15 = $[(n^2 + n)/2]$ = $[(5^2 + 5)/2]$ = 30/2 = 15; onto n = 21, then 21 = $[(n^2 + n)/2]$ = $[(6^2 + 6)/2]$ = 42/2 = 21; and the pattern continues...

In summary, the Positive Triangular Numbers are calculated as:

n = 0, then 0 = [(n² + n)/2] = [(0² + 0)/2] = 0/2 = 0;
n = 1, then 1 = [(n² + n)/2] = [(1² + 1)/2] = 2/2 = 1;
n = 3, then 3 = [(n² + n)/2] = [(2² + 2)/2] = 6/2 = 3;
n = 6, then 6 = [(n² + n)/2] = [(3² + 3)/2] = 12/2 = 6;
n = 10, then 10 = [(n² + n)/2] = [(4² + 4)/2] = 20/2 = 10;
n = 15, then 15 = [(n² + n)/2] = [(5² + 5)/2] = 30/2 = 15;
n = 21, then 21 = [(n² + n)/2] = [(6² + 6)/2] = 42/2 = 21; and the pattern continues...

The abovementioned Triangular Numbers can then be used to repetitively create all positive numbers in groups of 3 to illustrate that all positive numbers (positive integers) are Triangular Numbers.

For example:

n = 0, because, [0] + [0] + [0] = 0;
n = 1, because, [0] + [0] + [1] = 1;
n = 2, because, [0] + [1] + [1] = 2;
n = 3, because, [1] + [1] + [1] = 3;
n = 4, because, [0] + [1] + [3] = 4;
n = 5, because, [1] + [1] + [3] = 5;
n = 6, because, [0] + [3] + [3] = 6;
n = 7, because, [1] + [3] + [3] = 7;
n = 8, because, [1] + [1] + [6] = 8;
n = 9, because, [0] + [3] + [6] = 9;
n = 10, because, [0] + [0] + [10] = 10;
n = 11, because, [0] + [1] + [10] = 11;
n = 12, because, [1] + [1] + [10] = 12;
n = 13, because, [0] + [3] + [10] = 13;
n = 14, because, [1] + [3] + [10] = 14;
n = 15, because, [0] + [0] + [15] = 15;
n = 16, because, [0] + [1] + [15] = 16;
n = 17, because, [1] + [1] + [15] = 17;

n = 18, because, [0] + [3] + [15] = 18;
n = 19, because, [1] + [3] + [15] = 19;
n = 20, because, [0] + [10] + [10] = 20;
n = 21, because, [0] + [0] + [21] = 21;
n = 22, because, [0] + [1] + [21] = 22;
n = 23, because, [1] + [1] + [21] = 23;
n = 24, because, [0] + [3] + [21] = 24;
n = 25, because, [1] + [3] + [21] = 25;
n = 26, because, [1] + [10] + [15] = 26;
n = 27, because, [6] + [6] + [15] = 27;
n = 28, because, [0] + [0] + [28] = 28;
n = 29, because, [0] + [1] + [28] = 29;
n = 30, because, [0] + [15] + [15] = 30;
n = 31, because, [0] + [3] + [28] = 31;
n = 32, because, [1] + [3] + [28] = 32;
n = 33, because, [6] + [6] + [21] = 33;
n = 34, because, [0] + [6] + [28] = 34;
n = 35, because, [1] + [6] + [28] = 35;
n = 36, because, [0] + [0] + [36] = 36;
n = 37, because, [0] + [1] + [36] = 37;
n = 38, because, [1] + [1] + [36] = 38;
n = 39, because, [0] + [3] + [36] = 39;
n = 40, because, [6] + [6] + [28] = 40;
n = 41, because, [3] + [10] + [28] = 41;
n = 42, because, [0] + [6] + [36] = 42;
n = 43, because, [0] + [15] + [28] = 43;
n = 44, because, [6] + [10] + [28] = 44;
n = 45, because, [0] + [0] + [45] = 45; and the pattern continues...

A Summative Graph of Tripositive Trichotomous Upright Right Triangle Integers

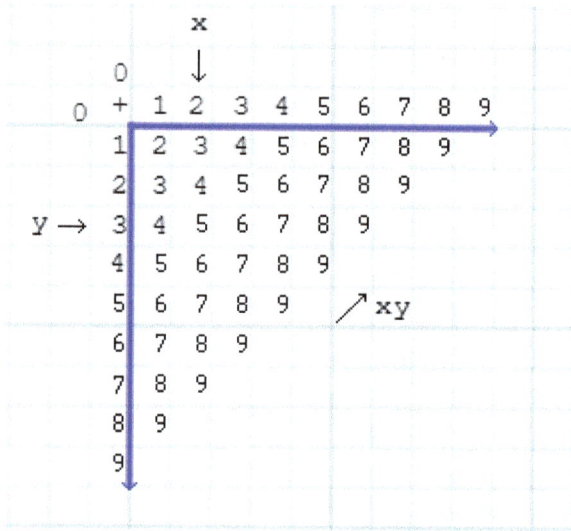

The Triological Science "Trifold Trichotomous Mathematical Operations"

Triangle Numbers have "Trifold Trichotomous Mathematical Operations" that parallel the Triological Science of Triology that observes nature's subatomic structure of the atom: Positive [as the Proton], Negative [as the Neutron], and Neutral [as the Electron]. The "Trifold Trichotomous Mathematical Operations" have a trichotomy of 3 "▽ Trichotomous Mathematical Operations (▽ Multiplication, ▽ Addition, and ▽ Subtraction) that in summation together create the final "▽Trichotomous Summative Harmonic Equation" that fields the 3-4-5-6 Golden Upright Right Triangle that has multiple definitive trichotomous mathematical qualities, characteristics, and elements. All of the aforementioned is illustrated, calculated, and represented graphically in the illustrated graphs and equations that follow.

The Meronymic (meaning "parts") Mathematics Operations of the "▽ Trichotomous Mathematical Operation" of the "▽ Trichotomous Summative Harmonic Equation and Operation" and the Holonymic (meaning "whole") Final "▽ Trichotomous Summative Harmonic Equation and Operation"

The "▽ Trichotomous Meronymic Multiplication Mathematical Equation and Operation" and its Geometric Illustration:

$$\nabla[x \cdot y] =$$

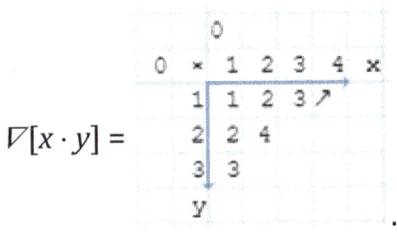

The "▽Trichotomous Meronymic Summation Mathematical Equation and Operation" and its Geometric Illustration:

$$\nabla[x + y] =$$

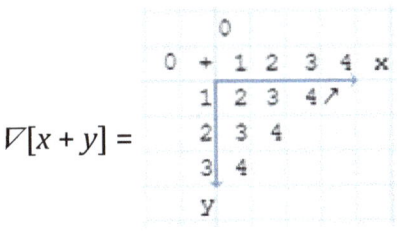

The "∇Trichotomous Meronymic Subtraction Mathematical Equation and Operation" and its Geometric Illustration:

$$\nabla[x-y] = \begin{array}{c|cccc} 0 & & & & \\ 0 & - & 1 & 2 & 3 & 4 \ \ x \\ 1 & 0 & 1 & 2\nearrow \\ 2 & -1 & 0 \\ 3 & -2 \\ y \end{array}$$

The Final "∇ Trichotomous Holonymic Summative Harmonic Operation" and its Geometric Illustration:

$$\nabla\Big[[x\cdot y] + [x+y] + [x-y]\Big] = \begin{array}{c|cccc} 0 & & & & \\ 0 \ [\,] & 1 & 2 & 3 & 4 \ \ x \\ 1 & 3 & 6 & 9\nearrow \\ 2 & 4 & 8\cdot \\ 3 & 5\cdot \\ y \end{array}$$

All of the ∇ Trichotomous Mathematical Operations in a Geometric Illustration exhibiting the holonymic sequential nature of the Equations that explain, define, lead to, and are a part of the 3-4-5-6 Golden Upright Right Triangle:

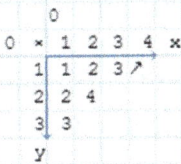

[x · y] =

```
     0
 0  ×  1  2  3  4  x
 1|    1  2  3↗
 2|    2  4
 3|    3
    y
```

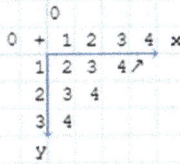

[x + y] =

```
     0
 0  +  1  2  3  4  x
 1|    2  3  4↗
 2|    3  4
 3|    4
    y
```

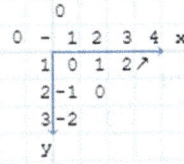

[x - y] =

```
     0
 0  -  1  2  3  4  x
 1|    0  1  2↗
 2|   -1  0
 3|   -2
    y
```

[x · y] + [x + y] + [x - y] =

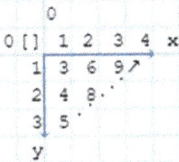

```
     0
 0  []  1  2  3  4  x
 1|     3  6  9↗
 2|     4  8 ·
 3|     5 ·
    y
```

=

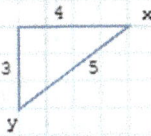

```
        4
              x
3 |    \
       |  \  5
       |____\
    y
```

=

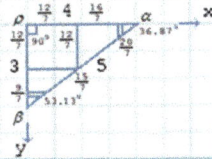

67

Why the 3-4-5-6 Upright Right Triangle is "Golden"

The uniqueness of the equality of the 3 by 3 Table with the 4 by 3 Table to create the 3-4-5-6 Golden Upright Right Triangle is illustrated in the following graphic and mathematical equations and formulae in the section that follows immediately after expressed using Trioengineering Notation:

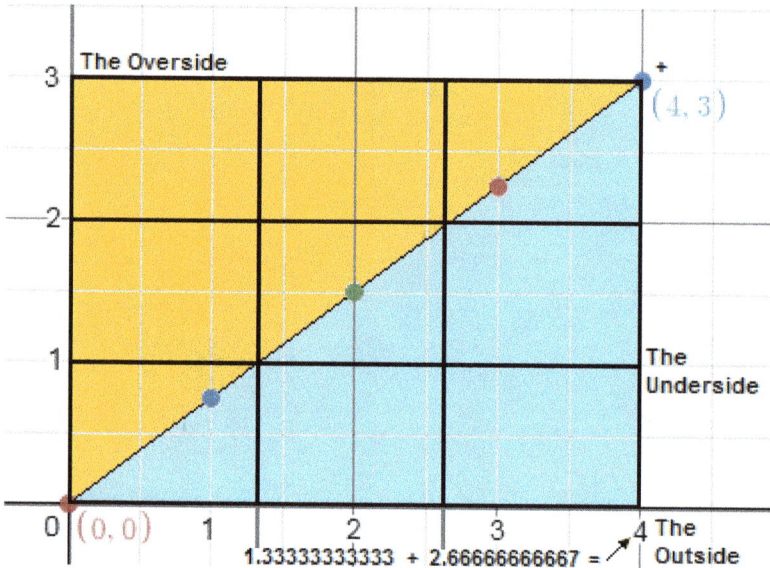

The uniquely inherent characteristics as unit measurements of the 3-4-5-6 Trichotomous Upright Right Triangle are what makes it "Golden". The calculative characteristics allow for the 3-4-5-6 Trichotomous Upright Right Triangle to create a triune trichotomous unity between three of the special and remarkable mathematical constants that can be found in and derived from nature.

They are the "Trichotomous Upright Right Triangle Pi" = " $\sqrt{\pi}$ " = " $\pi_{[\sqrt{}]}$ " (which is the difference the circumference of a circle and its diameter);

The "Trichotomous Upright Right Triangle Natural Logarithm" = " ∇e " = " $e_{[\nabla]}$ " (which is the); and the "Trichotomous Up Right Triangle Golden Ratio" = " $\nabla\Phi$ " = " $\Phi_{[\nabla]}$ " (which is the final value of a given line segment split into two pieces of different lengths so that the longer segment is equal to double the shorter segment). All three of the aforementioned are equal within the confines of the Trichotomous Upright Right Triangle. The "Golden" part of the name of the 3-4-5-6 Upright Right Triangle comes from the "Golden" in the "Golden Triangle Ratio" that is exactly the same as original "Golden Ratio" and is a part of the entire 3-4-5-6 Golden Upright Right Triangle. The precise calculations for each of the three trichotomous calculations with each of their respective original calculations are as follows:

1.) "Trichotomous Upright Right Triangle Pi" = " $\nabla\pi$ " = " $\pi_{[\nabla]}$ " = "The Ideal Trichotomous Upright Right Triangle Pi" = $\pi_{[\triangleright]}$ = $\left[\overline{x}_{[\triangleright]} \div \sqrt{\Phi}\right]$ = 3.144605511029693... (which is very close to the traditional numerical value of "pi") $\cong 3.141592654$;

2.) "Trichotomous Upright Right Triangle Natural Logarithm" = " ∇e " = " $e_{[\nabla]}$ " = "The Ideal Trichotomous Upright Right Triangle Natural Logarithm e " = $e_{[\triangleright]}$ = $\left[\text{Inscribed Insquare Side } (s) + \nabla abc\right]$ = 2.714285714285714285... (which is very close to the traditional numerical value of "e") $\cong 2.718281828459045...$; and

3.) "Up Right Triangle Golden Ratio" = " $\nabla \overset{\phi}{\uparrow}$ " = " $\overset{\phi}{\uparrow}_{[\nabla]}$ " = "The Ideal Trichotomous Upright Right Triangle Phi" =

$$\overset{\phi}{\uparrow}_{[\triangleright]} = \left[1 + \sqrt{ b \left[\sqrt{ 1 + \left[\frac{a}{b} \right]^2 } \right] } \right] \div 2 = 1.618033988749895\ldots$$

(which is exactly identical to the traditional numerical value of "phi" = "ϕ") $\equiv 1.618033988749895\ldots$

Graphically all of the aforementioned mathematical equations and their equality as identities associated in and as elemental foundational characteristics of the "3-4-5-6 Golden Upright Triangle" can be represented in the following definitive illustration:

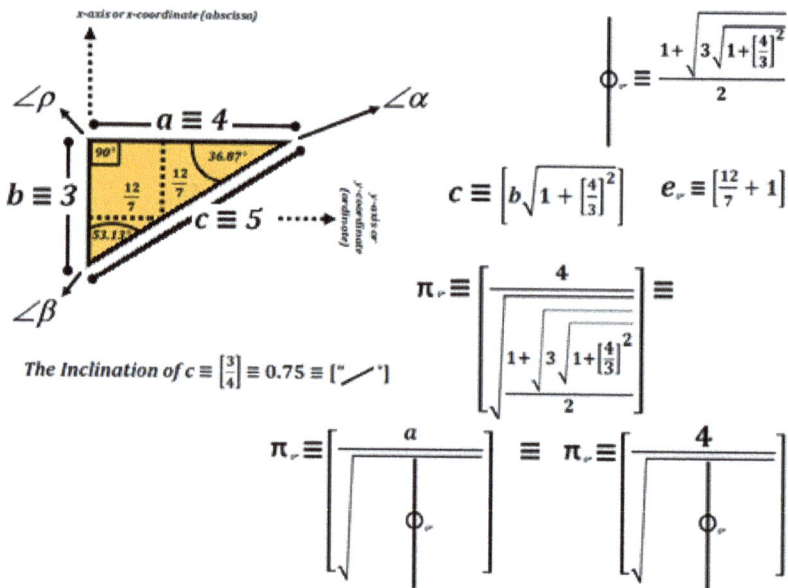

The Inclination of $c \equiv \left[\frac{3}{4}\right] \equiv 0.75 \equiv [\text{"}\diagup\text{"}]$

TRICHOMETRY © *The Study of the Geometrics of the 3-4-5-6 Golden Upright Right Triangle in Cartesian Coordinates.* Osler Studios Incorporated ™ © Copyright 2022 All Rights Reserved.

The Extracting of $\sqrt{abc}$ from the Visualus Isometric Cuboid

Reducing **xyz** from three coordinates into two coordinates that consist of **x** and **y**:

$$\frac{[xyz]}{[z]} = \frac{[x][y][z]}{[z]} = \frac{[x][y][\cancel{z}]}{[\cancel{z}]} = \frac{[x][y]}{1} = [x][y] = [xy] \text{ as}$$

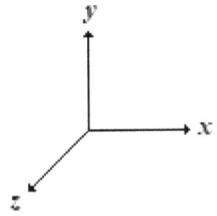

into $\llcorner$ thus,

$$[abc] = [a][b][c] \text{ into: } \frac{[abc]}{[c]} = \frac{[a][b][c]}{[c]} = \frac{[a][b][\cancel{c}]}{[\cancel{c}]} = \frac{[a][b]}{1} = [a][b] = [ab],$$

therefore **[xy] = [ab]**,

just as **[xyz] = [abc]**.

This then yields the following:

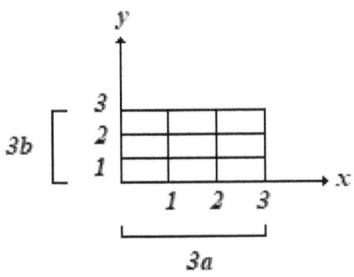

Where, the size (or "magnitude") on the x-axis = 4/3 per section (as a distance or measurable length) of each cell. Conversely, the size (or "magnitude") on the y-axis = 1 per section (as a distance or measurable height) of each cell. This can be exhibited graphically in the following manner:

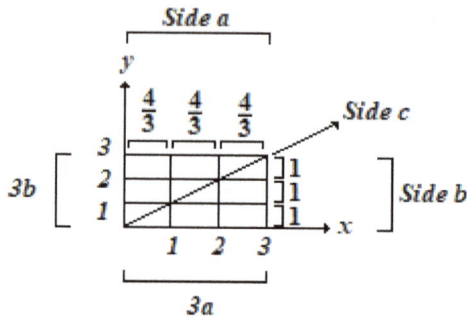

According to the model displayed above: $\frac{4}{3} = 1.333...$, Note: $[1.333... \times 3] = 4$. The 3 by 3 Table Format is composed of 2 Right Triangles one is upright and the other is inverted. This is illustrated in the following manner:

Note: 2 Right Triangles

Upright = [upr] 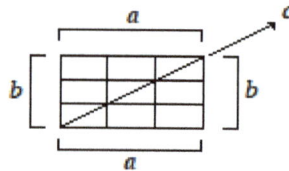 Inverted = [inv]

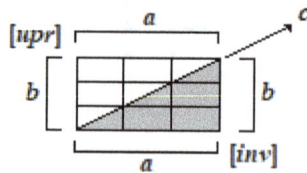

Note: **[upr]** = ("Upright") indicated by the unshaded Trichotomous Upright Right Triangle and opposite of it is **[inv]** = ("Inverted") indicated by the shaded light-gray area, together the two areas create the Standard 3 by 3 Table that is in tri-coordinates the front face of the Visualus Isometric Cuboid. The measurements of the 3-4-5-6 Golden Upright Right Triangle = "∇abc" are: *Side a* = 4; *Side b* = 3; *Side* c = 5; with a final *Area* of "A" = 6, respectively.

The Geometric Explanation of the "∇ Right Conversion" to create the "∇ Right Triangle Trichotomous Equation Characteristics of Side *c*"

The key to understanding all of the Triological Sciences begins with comprehending their foundation which is the 3-4-5-6 Golden Upright Right Triangle. To ideally grasp all of the meanings and subtle nuances of the Golden Upright Right Triangle one must first understand that it has its origin in the mathematics of Visualus and its foundation which is grounded in the Isometric Cuboid. The Isometric Cuboid is the visual manifestation of the Rectilinear (also known as Linear) Model of Instructional Systems Design. The Front Face of the Isometric Cuboid is where the 3-4-5-6 Golden Upright Right Triangle is derived. The Front Face is a 3 by 3 Table composed of 9 cells that make up the 9 sections of the original Isometric Cuboid. The 3 by 3 Table must be converted or transformed into a 4 by 3 Table of 12 equally proportioned 4 by 3 " 1 by 1 Squares" to begin constructing the three side unit measurements of the Golden Upright Right Triangle. This is referred to as mathematically "Squaring the Cuboid" as it is a mathematical geometric transformation from the Cuboid to a 4 by 3 Table to derive the Golden Upright Right Triangle (alternatively the transformation from the 4 by 3 Table into the full Isometric Cuboid is referred to as "Cubing to the 1 by 1 Square" or more simply called "Cubing the Square"). All of this is possible due to the unique relationship that exists between Sides *a* and *b* respectively of the 3-4-5-6 Golden Upright Right Triangle.

The Golden Upright Right Triangle "Side *a*" always has a unit measure of 4 and "Side *b*" always has a unit measure of 3. The slope/incline/acclivity of the Golden Upright Right Triangle when it sections the original 3 by 3 Table into two halves is [$\sqrt{m}$] = 1 because, [Side *a* ÷ Side *b*] = 3 ÷ 3 = 1. However, when the "Cubing the Square" transformation into the 4 by 3 Table occurs the slope/incline/acclivity of the Golden Upright Right Triangle now change and becomes [$\sqrt{m}$] = 0.75 because, [Side *a* ÷ Side *b*] = 4 ÷ 3 = 0.75. Thus, the same shape can now be represented as two equal ratios to produce the same outcomes in a variety of ways. This allows for the unique features of the original shape and its mathematics to be extended into multiple useful solutions and applications that have great utility in variety of circumstances and makes the 3-4-5-6 Trichotomous Upright Right Triangle a unilaterally useful tool. This is also one of the unique features that makes the 3-4-5-6 Trichotomous Upright Right Triangle "Golden" and initially why it is referred to as the "Golden Upright Right Triangle". The mathematical geometrics of the "Squaring the Cuboid" transformation is illustrated below and occurs in the following manner:

From: into: and converting from:

into: for V. The mathematics of the above can be presented graphically as:

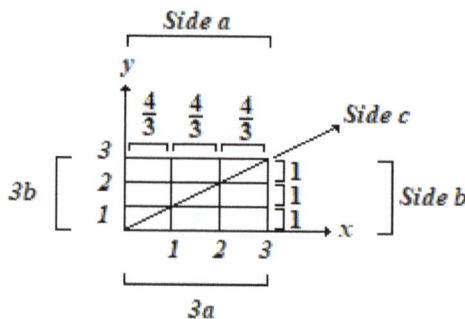

TRICHOMETRY $\overset{TM}{\circ}$ *The Study of the Geometrics of the 3-4-5-6 Golden Upright Right Triangle in Cartesian Coordinates.* Osler Studios Incorporated ™ © Copyright 2022 All Rights Reserved.

Transforms into the following on a 1 by 1 Square Graph:

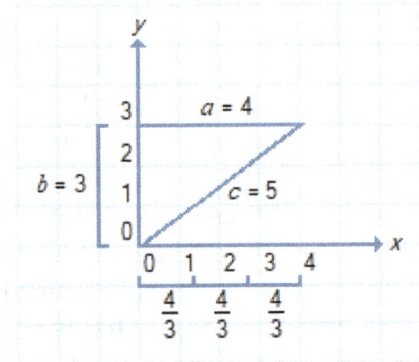

On the graph above: $x = (4, 0)$; and $y = $ Side $b = (0, 3)$.

Explaining how Side $a = 4$ from the "rectangularity" of the 3 by 3 Table (that is the Visualus "Front Face" of the Isometric Cuboid and also the Tri^2 Test 3 by 3 Table Format) can be further illuminated via graphical definition. By defining the equality of the 3 by 3 Table with the 4 by 3 Table by illustrating the equality of the distance of the respective cells, a greater understanding of the conversion and transformation of the 3 by 3 into the 4 by 3 for the unique purposes of measurement can be attained. The illustration below provides the definitive scale measurements of both Tables for comprehension of conversion and transformation in the following manner:

In the above "Table Equity Conversion and Transformation Illustration" equity is illustrated by the rectangular "3 by 3 Front Face" as one unit on the *y*-axis that has 3 units of 1, while the *x*-axis is equal to 4/3rds = 1.333... into 4 units of 1 on the final 4 by 3 Table. Thus, for the purposes of conversion and transformation that will yield the Pythagorean theorem, extensive general area field dynamics that have greater volumetrics, and Triophysics definitive areas that have broad utility in learning, Triostatistical analytics, and measurable self-growth. Therefore, the 3 by 3 Table is converted and transformed into the 4 by 3 Table in terms of singular cell rectangle units for the purposes of equality from a rectangle into a rectangle that is a 1 by 1 square. This is done by each of the individual cells (as individual singular units on the entire 3 by 3 Table) being sub-divided into Table cell units of a 4 by 3 structure with a total of twelve "1 by 1 Squares" thereby increasing the overall number of cells from 9 into 12 with side lengths changed from 1333... into units of 1. Mathematically, the ratios for the cells of the 4 by 3 Table with the 3 by 3 Table is a difference of 1/3rd that is equal to 1, written as follows:

$$\frac{x}{y} = \frac{4}{3} - \frac{1}{3} = \frac{3}{3} = 1.$$

The Conversion and Transformation from the 3 by 3 Table into the 4 by 3 Table

Graphically the conversion and transformation of the 3 by 3 into the 4 by 3 is illustrated as follows:

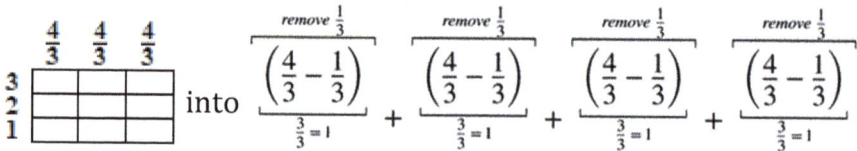

Which then via mathematical geometric conversion and transformation creates the following:

y = 3 $\begin{array}{c}3\\2\\1\end{array}$ [grid] = $\begin{array}{c}3\\2\\1\end{array}$ [grid]   y = 3

    1 2 3   1 2 3 4

$$x = 3 \qquad\qquad x = 4$$

For each $x = 4/3$,   For each $x = 1$

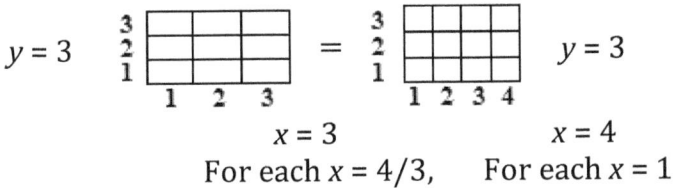

Therefore, the following is the explicative mathematical equation for the above geometric model as a mathematical equation.

The "Side a Conversion and Transformation from the 3 by 3 Table into the 4 by 3 Table" has the following equation:

$$\nabla\text{Side } a = [x_{1\ldots3} - 1/3] + [x_4 = 1/3(3)] = [1 + 1 + 1] + [1] = 4.$$

Where,

$$x_{1\ldots3} = 4/3 \text{ for } x_1, x_2, \text{ and } x_3.$$

Thus, the ∇ Side a for ∇abc Equation is the same as the "Side a Conversion and Transformation Equation" and is:

$$\nabla\text{Side } a = [x_{1\ldots3} - 1/3] + [x_4 = 1/3(3)] = [1 + 1 + 1] + [1] = 4.$$

In deference to the above, line xy = Side a = 4; Side b = 3 = The distance of Side a from the x-axis. All of this points on line xy satisfy the condition $y = b$ or y = Side b. Thus, if p(x, y) is any point on line xy, then $y = b$. As such, the equation of Side a as "line a" abc is a rectilinear straight line parallel to the x-axis is $y = b$. The equation of the x-axis for $\nabla abc = $ Tri$_{[abc]}$ is clearly y-axis Side b = 3 at the point where Side a = 4 is created as line xy, where p(x, y) is any point on the line xy that the endpoint is 4. Thus, the Final Triangle Equation of the Parallel for Side a for $\nabla abc = $ Tri$_{[abc]}$ is: y = Side b = 3 or $y = b$ = 3. The above graph can be explained and defined numerically (from the abovementioned "Table Equity Conversion and Transformation Illustration") in the following manner:

$$\frac{4}{3} \quad \frac{4}{3} \quad \frac{4}{3}$$

0 1 2 3

0 1 2 3 4

$$\frac{3}{3} \quad \frac{3}{3} \quad \frac{3}{3} \quad \frac{3}{3}$$
$$= \quad = \quad = \quad =$$
$$1 \quad 1 \quad 1 \quad 1$$

The above can be graphically defined this is provided:

The 3 by 3 Table = transforms into the 4 by 3 Table = that yields the following:

That produces the two Triangles = into the two oppositional Triangles = , the final result being the following:

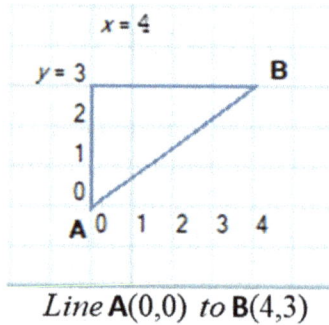

Line **A**(0,0) *to* **B**(4,3)

Thus, 3 by 3 Table Area as the Rectangle Area = A = $\sqrt{}abc$ + Δcab =

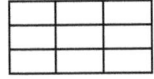

. By removing the inverse of $\sqrt{}abc$ as Δcab via subtraction in the following mathematical operation:

$\sqrt{}abc$ + Δcab
$\underline{\quad - \Delta cab}$ =
$\sqrt{}abc$

 Where, $\sqrt{}abc$ = and $\left.\begin{array}{l} a = 4; \\ b = 3; \\ c = 5; \text{ and} \\ A = 6. \end{array}\right\}$

Noting that $\sqrt{}abc$ is the "3-4-5-6 Golden Upright Right Triangle" (or "3-4-5-6 GURT"). All of the aforementioned together is the "Foundational Theorem of Triological Science" (also called "The Foundational Theorem of the Triological Sciences"), formulaically written graphically as:

 $= \dfrac{[xyz]}{[z]} = [xy] =$ $= [\sqrt{}abc + \Delta cab] - \Delta cab$ $\left.\begin{array}{l} a = 4; \\ b = 3; \\ c = 5; \text{ and} \\ A = 6. \end{array}\right\}$

Expressed without geometrics (with full Trines represented) as:

$\dfrac{[xyz]}{[z]} = [xy] = [\sqrt{}abc + \Delta cab] - \Delta cab = \sqrt{}abc$ $\left.\begin{array}{l} a = 4; \\ b = 3; \\ c = 5; \text{ and} \\ A = 6. \end{array}\right\}$

Rewritten numerically as:

$$[9xy] \equiv [9ab] \div \tfrac{1}{2}_{[inv]} \equiv \sqrt{abc} \quad \left.\begin{matrix} a = 4; \\ b = 3; \\ c = 5; \text{ and} \\ A = 6. \end{matrix}\right\}$$

This is defined in a mathematical geometric identity (that uses the mathematic "identical to" symbol of "≡" as follows:

$$[9xy] \equiv [9ab] \div \Delta \equiv \sqrt{abc}.$$

Where the above elements of the identity have the following definitions:

$[9xy]$ = The precise Cartesian Coordinates measurements of

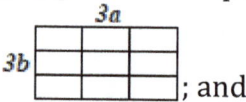

; and

$$[9ab] \div \tfrac{1}{2}_{[inv]} = \quad - \Delta cab.$$

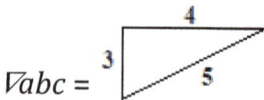

Where the above elements of the equations presented immediately above have the following further definitions:

$\tfrac{1}{2}_{[inv]} = \Delta cab;$

$\sqrt{abc} =$ that has an Area of "A", where A = 6; and

$\Delta = $ $= \frac{1}{2}_{[inv]} = 0.5_{[inv]}$ (where, "[*inv*]" = inverted which equates to the inverse (or the "upside down and reversed opposite") of the initially presented Trichotomous Upright Right Triangle, that when placed together with original Trichotomous Upright Right Triangle create the original 3 by 3 Table).

This can be rewritten with mathematical parsimony as "The $\sqrt{abc}$ Foundational Theorem of Triological Science (also called "The Triological Science Foundational Theorem of $\sqrt{abc}$") as:

$$\frac{[9ab]}{0.5_{[inv]}} = \sqrt{abc} \equiv$$ $\begin{cases} a = 4; \\ b = 3; \\ c = 5; \text{ and} \\ A = 6. \end{cases}$

In this manner all of the characteristics and traits of the 3-4-5-6 Golden Upright Right Triangle are inherited by the Trine abc as the representative mathematical symbol of the larger more detailed GURT model as a prime example of mathematical parsimony. Formulaically this can be written via a mathematical equation in the following manner:

$$\sqrt{abc} = $$ $= $ $= \frac{1}{2}_{[upr]} = 0.5_{[upr]}$ (where, "[*upr*]" = the Trichotomous Upright Right Triangle).

The "∇ Right Triangle y–Intercept Equation" of the "∇ Right Triangle = $\sqrt{abc}$"

The "∇ Right Triangle Inclination of Side *c*" as the "∇ Right Triangle Acclivity of Side *c*" is represented by the following equation: $[\nabla m \cdot \nabla x] = [\nabla mx], b = 0, and [a/b = \frac{3}{4}]$. This then allows for the "∇ Right Triangle y–Intercept Equation" to be written as: "$\nabla y = [\nabla mx] + b$".

To create the GURT the first step requires the Front Face of the Visualus Isometric Cuboid to be transformed through the three Triological Science "Trichotomous Trioengineering Transformations" that are: (1.) "$\sqrt{}$Right Triangle Trioengineering Intercalation"; (2.) "$\sqrt{}$Right Triangle Trioengineering Inculcation"; and (3.) "$\sqrt{}$ Right Triangle Trioengineering Interpolation" for the Conversion of the Standard 3 by 3 Table into a 4by 3 Table for the creation of the Golden Upright Right Triangle with measurements that are: Side $a = 4$, Side $b = 3$, Side $c = 5$, and Area A = 6.

Mathematically, the "$\sqrt{}$ Right Triangle y–Intercept Equation" is calculated in the following manner: $\sqrt{}y = [¾ \cdot 4] + 0$, therefore, $\sqrt{}y = 3$, thusly, $\sqrt{}m$ = " $\sqrt{}$Right Triangle Slope" = " $\sqrt{}$Right Triangle Inclination" = " $\sqrt{}$Right Triangle Acclivity" = 0.75 the full definition of $\sqrt{}m$ is: $\sqrt{}m$ = "$\sqrt{}$Right Triangle Slope" = " $\sqrt{}$Right Triangle Inclination" = " $\sqrt{}$Right Triangle Acclivity" $= \frac{y_2 - y_1}{x_2 - x_1} = \frac{3-0}{4-0} = \frac{3}{4} = 0.75$. The following graphical illustration is true according to the " $\sqrt{}$Right Triangular Interpolation Equation of $\sqrt{}abc$" for the inclination and acclivity of Side c:

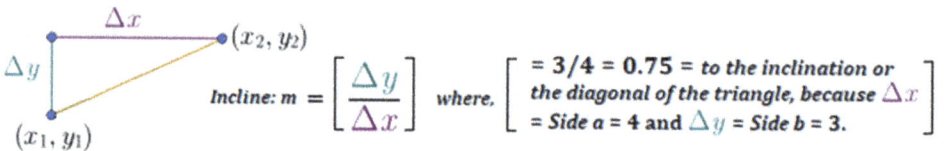

Defining and Explaining the Triological Science Trichotomous via the following Three Trichotomous $\sqrt{}$ Right Triological Trioengineering that are:

"The $\sqrt{}$Right Triangle Intercalation of $\sqrt{}abc$";
"The $\sqrt{}$Right Triangle Inculcation of $\sqrt{}abc$"; and
"The $\sqrt{}$Right Triangle Interpolation of $\sqrt{}abc$".

The Triological Science Trioengineering of $\sqrt{abc}$ occurs via the trichotomous triune calculative characteristics has three identified distinctive areas that most accurately define the overall uniqueness, utility, and usefulness of the 3-4-5-6 Golden Upright Right Triangle. The three calculative characteristics are viable in that they have interior, exterior, and ulterior measurements that are defined as: (1.) "$\sqrt{}$Right Triangle Intercalation" as the outside cyclical model of $\sqrt{abc}$; (2.) "$\sqrt{}$ Right Triangle Inculcation" as the inside holistic ideation and conceptualization within $\sqrt{abc}$; and (3.) "$\sqrt{}$Right Triangle Interpolation" as the .

The Geometric Explanation of the "$\sqrt{}$Right Triangle Intercalation Equation of $\sqrt{abc}$"

Illustrating the "external" (or "exterior") characteristics of the 3-4-5-6 GURT according to the "$\sqrt{}$Triangle Intercalation Equation of $\sqrt{abc}$". The "$\sqrt{}$ Triangle Intercalation Equation" involves the outside of the Triangular Equation Modeling [TEM] as first presented by the author in a 2017 published research paper entitled, "Triangular Equation Modeling [TEM]: The Base Operation, Basic Rationale, and Foundational Logic that is the Basis for "Research Architectural Metrics" that are Essential to the Planning, Scope, and Schema of Trichotomous Research Designs" in the i-manager's July–September Journal on Mathematics.

The base [TEM] can be illustrated as a whole as follows (Osler, 2017 in the published research paper entitled, "Triangular Equation Modeling [TEM]: The Base Operation, Basic Rationale, and Foundational Logic that is the Basis for "Research Architectural Metrics" that are Essential to the Planning, Scope, and Schema of Trichotomous Research Designs" that originally appeared in the i-manager's July–September Journal on Mathematics):

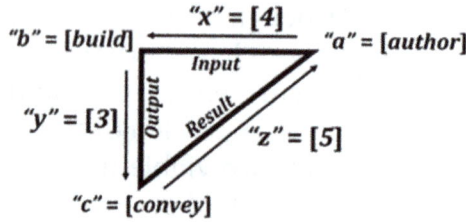

"b" = [build] ← "x" = [4] → "a" = [author]
Input
Output
Result
"y" = [3]
"z" = [5]
"c" = [convey]

The [TEM] as the Algorithmic Triangular Model (as presented by the author in the published 2021 research paper entitled, "Algorithmic Triangulation Metrics for Innovative Data Transformation").

Defining the Application Process of the Tri–Squared Test" as it in appeared in the April–June i-manager's Journal on Mathematics) is conveyed in the following illustration that illustrates "a trichotomy within a trichotomy" (Osler, 2021) as follows:

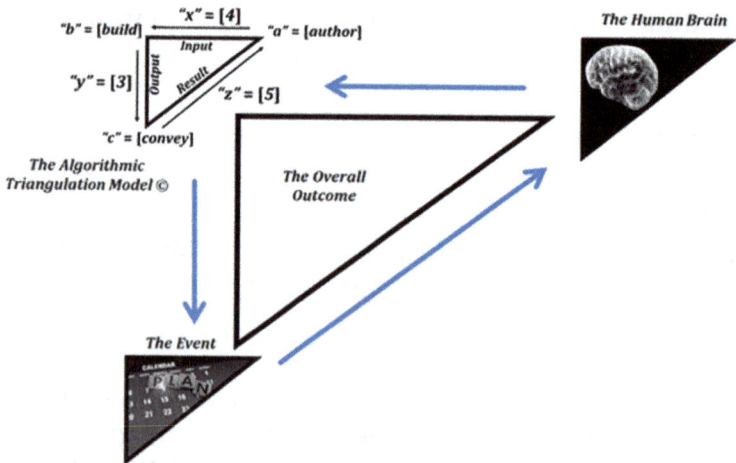

The Algorithmic Triangulation Model ©

The Overall Outcome

The Human Brain

The Event

The [TEM] utilizes the 3-4-5-6 GURT as a trichotomously explicative model to illustrate the unique relationships that exists between, within, and throughout a plethora of algorithms, axioms, concepts, details, ideas, identifications, innovations, inventions, investigations, measurements, principles, solutions, and theories.

When used to display the exterior as the outside of the [TEM] then for example the following approach is used:

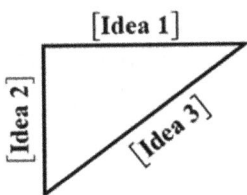

The [TEM] can be further defined contextually in terms of mathematical definitions in the following series of graphics and equations:

The Trichotomous Upright Right Triangle Testing Model Using [TEM] to more accurately describe the Tri–Squared Test and the origins of the Trine Symbol as Originally Published in 2012, 2017, and 2021

The Algorithmic Model of Triangulation (Osler, 2013 as it originally appeared in the referred publication by the author entitled, "Algorithmic Triangulation Metrics for Innovative Data Transformation: Defining the Application Process of the Tri–Squared Test" in the April–June Journal on Mathematics) is of the form:

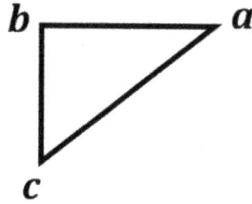

Where,

Vertex a = $\nabla\angle a$ = form "Vector $\nabla[a{\rightarrow}b]$" as " $\nabla[\overrightarrow{ab}]$" = "**authoring**" = a ratio of a measurement metric of 4 (used to convey information) = The Initial Tri–Squared Instrument Design;

Vertex b = $\nabla\llcorner b$ = = form "Vector $\nabla[b{\rightarrow}c]$" as " $\nabla[\overrightarrow{bc}]$" = "**building**" = a ratio of a measurement metric of 3 (as the Triostatistics: Tri–Squared Test) = The Tri–Squared Qualitative Instrument Responses which will be analyzed via Tri–Squared Analysis; and

Vertex c = $\nabla\angle c$ = = form "Vector $\nabla[c{\rightarrow}a]$" as " $\nabla[\overrightarrow{ca}]$" = "**conveying**" = The Final Tri–Squared Test Outcomes in a Quantitative Report.

Thus, the Triangulation Model is symbolized by a Right Triangle written in graphical illustrative format as the following image: " ∇ " = the italicized nabla as the Trioengineering Notation Trine " ∇ " that has appeared earlier in this narrative. This symbol is called the "Trine" (meaning a group of three) is written mathematically as " ∇ = abc" = " ∇abc" and is simplified into the mathematic geometric expression: ∇abc (meaning "Triangulation Model abc" or more simply "Trine abc" which is the original definition as first published in 2013). The adapted definitions are presented here from that original publication in deference to the Triological Sciences as follows:

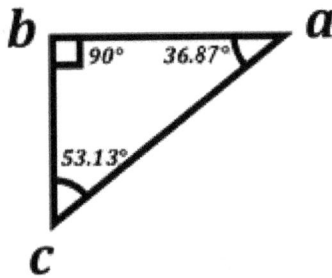

The angles have the following angular measurements in degrees that measure radially as: ∠*a* = 36.87 degrees, ∟*b* = 90 degrees, and ∠*c* = 53.13 degrees respectively, that all add up to the standard 180° of any and all triangles, thus, in terms of the Trioengineering Notation Trine as the GURT is: [36.87° + 90° + 53.13° = 180°]. The connective lines between the Trichotomous Upright Right Triangle vertex points (i.e., the lines between points *a*, *b*, and *c* respectively are geometric "vectors" (lines with both magnitude [or "size"] and direction) making the model a systemic or cyclic process from the point of origin "∠*a*" back to the original point of origin which is also "∠*a*". This is illustrated in terms of Cartesian Coordinates as follows:

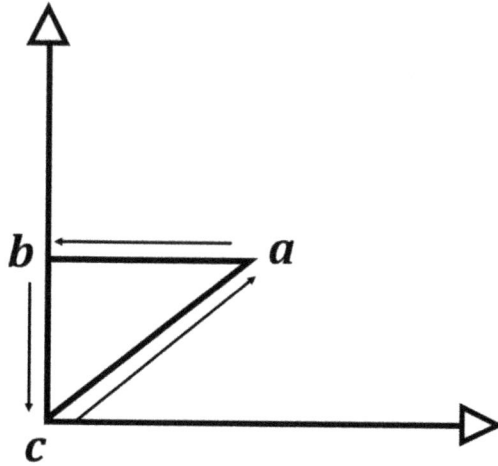

The model above is the beginning of the systemic cyclical methodology of [TEM]. In terms of Vectors, the Trichotomous Upright Right Triangle as the Algorithmic Model of Triangulation now becomes the Trioengineering: Triangulation Model Right Triangle ∇abc is equal to three vectors that illustrate the movement in direction and magnitude from one completed task into another. The entire process is both cyclical and sequential with a "Trine Vector Equation" (using the graphical Trine illustration image) written in Trioengineering Notation as:

$$\nabla = \nabla[\leftarrow x] \rightarrow \nabla[\downarrow y] \rightarrow \nabla[\nearrow z] = \nabla[\nearrow xy], \textit{ for } \text{"} \nabla[\overrightarrow{xy\dot{z}}]\text{"} \textit{ as } \text{"} \nabla[\overrightarrow{xy\dot{z}}]\text{"}$$
$$\textit{inbetween} \text{ "} \nabla\angle abc\text{"}$$

Defined as Trine = "Concentration of Vector x into Concentration of Vector y into Concentration of Vector z", which is simplified into a more standardized Trine Vector Equation form (using the graphical Trine illustration image) also written in Trioengineering Notation as:

$$\nabla = \nabla\overleftarrow{x} \rightarrow \nabla\overrightarrow{y} \rightarrow \nabla\overrightarrow{z}, \textit{ for } \text{"} \nabla\overrightarrow{xy\dot{z}}\text{"} \textit{ as } \text{"} \nabla\overrightarrow{xyz}\text{"} \textit{ inbetween } \text{"} \nabla\angle abc\text{"}$$

Where vectors in terms of Trioengineering Notation are: $\nabla\hat{x}$, $\nabla\hat{y}$, and $\nabla\hat{z}$. Where, "z" is not nor has any relation to the Cartesian Coordinate z-axis (also known as the "applicate") here, instead "z" is a recognized vector that is parallel to Side c for the purposes of illustrating the cyclical nature of the 3-4-5-6 Golden Upright Right Triangle as a model (as used in the Triostatistics [TEM] operation and mathematical model). Each of the indicated vectors respectively are indicated on the "Algorithmic Triangulation Data Model" written in a graphical illustration as:

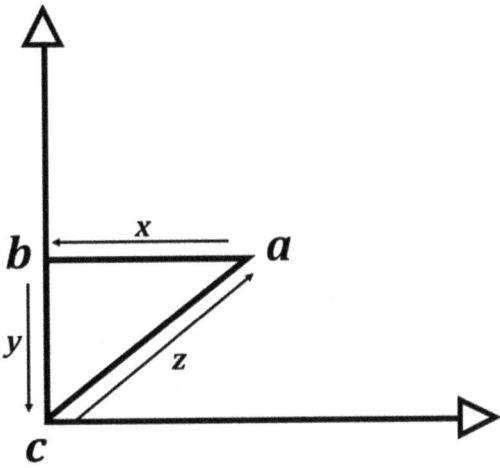

The previous illustration indicates that the standardized form of the Trioengineering Notation Trichotomous Vectors in the model above are: $\nabla\hat{x}$, $\nabla\hat{y}$, and $\nabla\hat{z}$ are also equivalent to the Triangle's Sides (a, b, and c) that are the sequential Cartesian Coordinates relative to the size and magnitude of the research engineering phases that sequentially connect the respective angles and report them using Trioengineering Notation as: $\nabla\angle a$, $\nabla\llcorner b$, and $\nabla\angle c$. The aforementioned are written in a sequential cycle illustrating the flow of the vectors in a counterclockwise direction opposite of the Cartesian Coordinates in the first quadrant indicating the more precise direction of arrows from right to left in a motion that can be expressed using Trioengineering Notation as follows:

$$\overleftarrow{\nabla x} = \overleftarrow{\nabla ab};$$
$$\overleftarrow{\nabla y} = \overleftarrow{\nabla bc}; \text{ and}$$
$$\overleftarrow{\nabla z} = \overleftarrow{\nabla ac}.$$

Where,

$\nabla z = \nabla xy$ with starting point = $(0, 0)$ to ending point $(4, 3)$ with a slope/incline/acclivity = 0.75.

The Complete Tri–Squared Analysis Algorithmic Triangulation Model as the 3-4-5-6 Trichotomous Upright Right Triangle (adapted and re-edited from the author's publication in 2013 entitled, "Algorithmic Triangulation Metrics for Innovative Data Transformation: Defining the Application Process of the Tri–Squared Test" in the April–June Journal on Mathematics)

The three numeric Vector Operational Phases of the Triangulation Model that now expresses the full scope of the operational parameters of the Triostatistics Tri–Squared Test are completely defined in the following manner (as presented earlier):

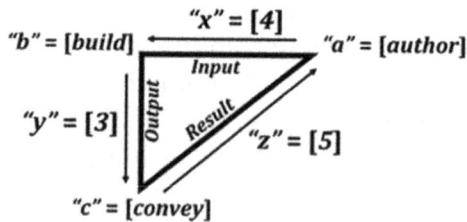

The entire Triangulation Model as a Research Engineering methodology begins with a breakdown of the Trine [∇] Operational Research Engineering Parameters and Geometric Vectors in the next section.

The Tri–Squared Triangulation Model Research Engineering Process (highlighting the Operational Parameters and Phases of the Tri–Squared Test) for the Triological Sciences

Geometric Vertex $\nabla a = \nabla \angle a =$ "**authoring**" = The Initial Tri–Squared Instrument Design = Operational Parameter " ∇a " = "author" = absolute value of $\nabla a =$ "modulus ∇a " = $|a|$ = "Trioengineering Notation Trine a " = $\nabla a =$ The creation of the Tri–Squared Inventive Investigative Instrument. This process can be seen in the following 3 by 3 cubic model which becomes the 3 by 3 rectangle table model (due to the 4/3rds conversion and transformation mathematics as mentioned in earlier sections of this narrative) of the Standard Tri–Squared Test Data Analysis Table (note: it becomes rectangular due to the top ratio author vector of "4" which represents the four initial sections of the Tri–Squared Test Triple-I instrument as the four Operational Phases of in-depth trichotomous research question data analysis:

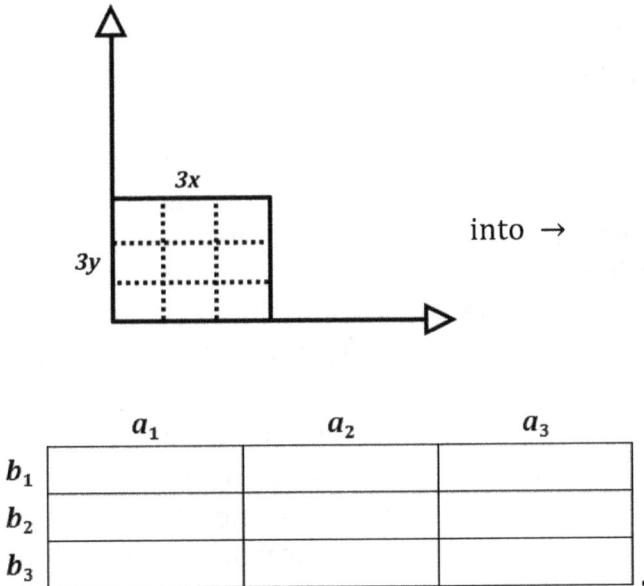

	a_1	a_2	a_3
b_1			
b_2			
b_3			

.

This in turn (written in Trioengineering Notation), leads into [→] vector ∇x = Geometric Vector ∇x = "← ∇x" = $\overleftrightarrow{\nabla x}$ = $\nabla \overleftarrow{ab}$ = The Initial Tri–Squared Instrument Construction = Operational Phases " ∇x" = absolute value of vector ∇x = "norm ∇x" = $||x||$ = "Trine x" = $\blacktriangledown x$ = The creation of the Tri–Squared Inventive Investigative Instrument = The Pythagorean Triple of the Triangulation Model = "4" = The **4 Phases of Tri–Squared Inventive Investigative Instrument Construction** as "**Trichotomous Psychometrics**" which is composed of 4 Operational Phases that proactively create the "Trioengineered Trichotomous Research Battery" for the Tri–Squared Test. It is this dynamically precise and meticulously detailed trichotomous research battery that is the research investigation "Triple–I" (or "Inventive Investigative Instrument") that is a "Triple Trichotomy" based upon the research investigation's initial three trichotomous research questions. The 4 Phases that attribute to the creation of the "Triple–I" are as follows:

1. a_0 = The Instrument Name (Asset Security, generally in the form of Copyright, or Creative Commons, and/or Trademark);

2. a_1 = Section One of the Research Instrument. Constructed from the first series of instrument items (a. through c.) derived from the research investigation questions as the Qualitative Trichotomous Categorical Variables (as the Initial Investigation Input Variables), evaluated via the Qualitative Trichotomous Outcomes (as the Resulting Outcome Output Variables = b_1, b_2, and b_3 respectively);

3. a_2 = Section Two of the Research Instrument. Constructed from the second series of instrument items (d. through f.) derived from the research investigation questions as the Qualitative Trichotomous Categorical Variables (as the secondary Investigation Input Variables), evaluated via the Qualitative Trichotomous Outcomes (as the Resulting Outcome Output Variables = b_1, b_2, and b_3 respectively); and

4. a_3 = Section Three of the Research Instrument. Constructed from the third series of instrument items (g. through i.) derived from the research investigation questions as the Qualitative Trichotomous Categorical Variables (as the tertiary Investigation Input Variables), evaluated via the Qualitative Trichotomous Outcomes (as the Resulting Outcome Output Variables = b_1, b_2, and b_3 respectively);

The following table provides the metrics for the construction of the Inventive Investigative Instrument following the parameters indicated in phases 1. through 4. of the first vector [$\overline{x} = \overleftarrow{ab}$ = 4] of the Triangulation Model:

	"Inventive Investigative Instrument" (or "Triple–I")		
$a_0 =$	—Name—		
	[Adding Asset Security as Copyright = © and/or Trademark = ™]		
$a_1 =$	Section 1. Research Question 1. [Trichotomous Research Battery: One] The First Series of Questions from the Qualitative Trichotomous Categorical Variables are listed below in Items a. through c.		
	Responses: [Select only one from the list.] ▶	b_1	b_2	b_3
	a. Item One based upon Research Question 1	☐	☐	☐
	b. Item Two based upon Research Question 1	☐	☐	☐
	c. Item Three based upon Research Question 1	☐	☐	☐
$a_2 =$	Section 2. Research Question 2. [Trichotomous Research Battery: Two] The Second Series of Questions from the Qualitative Trichotomous Categorical Variables are listed below in Items d. through f.		
	Responses: [Select only one from the list.] ▶	b_1	b_2	b_3
	d. Item Four based upon Research Question 2	☐	☐	☐
	e. Item Five based upon Research Question 2	☐	☐	☐
	f. Item Six based upon Research Question 2	☐	☐	☐
$a_3 =$	Section 3. Research Question 3. [Trichotomous Research Battery: Three] The Third and Final Series of Questions from the Qualitative Trichotomous Categorical Variables are listed below in Items g. through i.		
	Responses: [Select only one from the list.] ▶	b_1	b_2	b_3
	g. Item Seven based upon Research Question 3	☐	☐	☐
	h. Item Eight based upon Research Question 3	☐	☐	☐
	i. Item Nine based upon Research Question 3	☐	☐	☐

The mathematics of the "Trioengineered Tri–Squared Test Triple–I Template" illustration presented above are as follows (in terms of Trioengineering Notation equations) in the next series of equations.

The Complete Trioengineering Tri–Squared Test Triple–I as a "Psychometric Research Trichotomous Battery Template Equation" is:

$$\overset{3}{\underset{i=1}{V}}[a_0 + a_1 + a_2 + a_3].$$

The 4 Phase Construction Equations for the "Trioengineered Tri–Squared Test Triple–I Template":

$$\nabla a_0 = \nabla[\text{Tri–I}_{[name]} + \text{AS}_{[©; ™]}];$$
$$\nabla a_1 = \nabla[\text{TCV}_{[1]} = \text{TCV}_{1 = \text{Items: } [a....c.]} \rightarrow \text{TOV}_{\text{Tri} = [b_1...b_3]}];$$
$$\nabla a_2 = \nabla[\text{TCV}_{[2]} = \text{TCV}_{2 = \text{Items: } [d....f.]} \rightarrow \text{TOV}_{\text{Tri} = [b_1...b_3]}]; \text{ and}$$
$$\nabla a_3 = \nabla[\text{TCV}_{[3]} = \text{TCV}_{3 = \text{Items: } [g....i.]} \rightarrow \text{TOV}_{\text{Tri} = [b_1...b_3]}].$$

Where, each of the above are mathematically defined below by the individual 4 Phases of construction in the following manner:

The elements of **Phase One** are defined as follows:
∇a_0 = The Trioengineered Triple–I ("Inventive Investigative Instrument") Name;
$\nabla[\text{Tri–I}_{[name]}]$; = The Trioengineered "Triple–I Name" assigned to this psychometric instrument; and
$\nabla[\text{AS}_{[©; ™]}]$; = The Trioengineered Asset Security with Copyright = © and Trademark = ™.

The elements of **Phase Two** are defined as follows:
∇a_1 = The Trioengineered First Trichotomous Battery based upon the First Research Question as the First Trichotomous Categorical Variable (TCV) and its associated Sub-Questions as Items a., b., and c. respectively with the responses as the Trichotomous Outcome Variables (TOVs) that are b_1, b_2, and b_3 respectively;
$\nabla[\text{TCV}_{[1]}]$ = The Trioengineered Trichotomous Categorical Variable 1;
$\nabla[\text{TCV}_{1 = \text{Items: } [a....c.]}]$ = The Trioengineered (TCV 1) Sub-Questions as Items a., b., and c.;
$\nabla[\rightarrow]$ = The Trioengineered moving forward to Trichotomous Outcome Variables (TOVs); and

$\nabla[\text{TOV}_{\text{Tri} = [b_1 ... b_3]}]$ = The Trioengineered (TOVs) that are Trichotomous Outcomes b_1, b_2, and b_3.

The elements of **Phase Three** are defined as follows:
∇a_2 = The Trioengineered Second Trichotomous Battery based upon the Second Research Question as the Second Trichotomous Categorical Variable (TCV) and its associated Sub-Questions as Items d., e., and f. respectively with the responses as the Trichotomous Outcome Variables (TOVs) that are b_1, b_2, and b_3 respectively;

$\nabla[\text{TCV}_{[2]}]$ = The Trioengineered Trichotomous Categorical Variable 2;

$\nabla[\text{TCV}_{2 = \text{Items: } [d...f.]}]$ = The Trioengineered (TCV 2) Sub-Questions as Items d., e., and f.;

$\nabla[\rightarrow]$ = The Trioengineered moving forward to Trichotomous Outcome Variables (TOVs); and

$\nabla[\text{TOV}_{\text{Tri} = [b_1 ... b_3]}]$ = The Trioengineered (TOVs) that are Trichotomous Outcomes b_1, b_2, and b_3.

The elements of **Phase Four** are defined as follows:
∇a_3 = The Trioengineered Three Trichotomous Battery based upon the Three Research Question as the Three Trichotomous Categorical Variable (TCV) and its associated Sub-Questions as Items g., h., and i. respectively with the responses as the Trichotomous Outcome Variables (TOVs) that are b_1, b_2, and b_3 respectively;

$\nabla[\text{TCV}_{[3]}]$ = The Trioengineered Trichotomous Categorical Variable 3;

$\nabla[\text{TCV}_{3 = \text{Items: } [g...i.]}]$ = The Trioengineered (TCV 3) Sub-Questions as Items g., h., and i.;

$\nabla[\rightarrow]$ = The Trioengineered moving forward to Trichotomous Outcome Variables (TOVs); and

$\nabla[\text{TOV}_{\text{Tri} = [b_1 ... b_3]}]$ = The Trioengineered (TOVs) that are Trichotomous Outcomes b_1, b_2, and b_3.

The ∇ [TEM] for the "Trioengineered Tri–Squared Test Triple–I Template" illustration is represented geometrically in a Triostatistics [TEM] in the following manner:

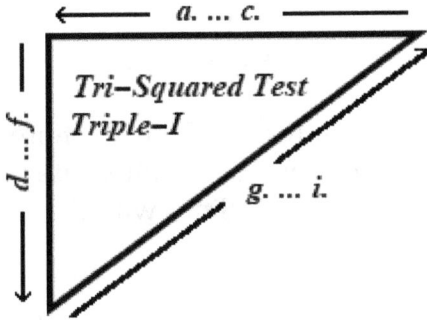

The **Mathematics of the Triangular Equation Modeling [TEM] in Trioengineering Notation Illustrating How the Triostatistics [TEM] comes out of the 3-4-5-6 Golden Upright Right Triangle as the ∇ [TEM] and How the In-Depth Measurements of the Golden Upright Right Triangle Actively and Geometrically Create the 3 by 3 Tri–Square Analysis Table**

The traditional Triostatistics "[TEM]" is equal to the "∇[TEM]" that emphasizes the connection to the "Trichotomous Upright Right Triangle" via the [TEM] originating from the 3-4-5-6 Golden Upright Right Triangle. The exterior of the [TEM] in deference to its use of the 3-4-5-6 Golden Upright Right Triangle expresses via Trioengineering Notation the Trichotomous Upright Right Triangle Perimeter as "∇P". The ∇P is mathematically expressed in the following series of mathematical formulae and calculations:

$$\nabla P = a + b + \sqrt{a^2 + b^2} = a + b + c$$

$$4 + 3 + \sqrt{4^2 + 3^2} = 7 + \sqrt{16 + 9} = 7 + \sqrt{25}$$

$$7 + 5 = 12, \text{ and}$$

$$[a + b + c] = [4 + 3 + 5] = 12.$$

Therefore, the full numerical Trichotomous Upright Right Triangle Perimeter = The " $\sqrt{}$[TEM] Perimeter" is:

$$\sqrt{}P = 12.$$

The Intercalation of $\sqrt{}abc$ in terms of the Trichotomous Upright Right Triangle Perimeter can be rewritten to illustrate the equity of the 3-4-5-6 Golden Upright Right Triangle Sides with the Cartesian Coordinates in the following manner:

$$\sqrt{}P = [a + b + c] = [x + y + z],$$

Where,

$\sqrt{}a = a = x;$
$\sqrt{}b = b = y;$ and
$\sqrt{}c = c = z.$

Holistically, the "Total Trichotomous Triune Calculative Characteristics of $\sqrt{}abc$" are expressed in a mathematical perimeter definition as follows:

$$\sqrt{}P \text{ of } \sqrt{}abc = \text{Side } a + \text{Side } b + \text{Side } c = 4 + 3 + 5 = 12,$$

Where,

$$\sqrt{}abc \left.\begin{array}{l} a = 4; \\ b = 3; \\ c = 5; \text{ and} \\ A = 6. \end{array}\right\}$$

The Geometric Explanation of the "▽Right Triangle Inculcation Equation of ▽abc"

Illustrating the "internal" (or "interior") characteristics of the 3-4-5-6 GURT according to the "▽Triangle Inculcation Equation of ▽**abc**". In this manner the ▽ Triangle Inculcation uses the internal holistic characteristic of the [TEM]. Where, ▽ Triangle Intercalation is concerned only with the interior of the [TEM] that illustrates the whole concept, idea, solution and/or problem under observation, investigation and/or study.

Thus, the ▽Triangle Intercalation as the [TEM] represented as the Algorithmic Triangular Model only displays the interior "Overall Outcome" as the holistic idea, concept, thought and/or problem exhibited as:

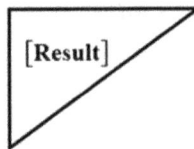

The "Result" as entered in the [TEM] above is also equal to the Area of the entire which is A = 6 as illustrated in the [TEM] below.

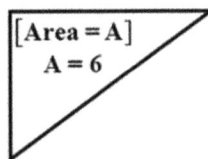

The interior of the [TEM] in deference to its use of the 3-4-5-6 Golden Upright Right Triangle expresses the Trichotomous Upright Right Triangle Area as "▽A". The ▽A is mathematically expressed in the following series of mathematical formulae and calculations:

$$\nabla A = \frac{1}{2}ab = \frac{ab}{2} = \frac{[\text{Side } a][\text{Side } b]}{2} = \frac{4 \cdot 3}{2} = \frac{12}{2} = 6$$

therefore,

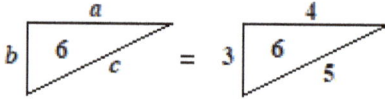

Where, $\nabla A = A = \text{Area} = 6$.

Therefore, the full numerical Trichotomous Upright Right Triangle Area = The " ∇[TEM] Area" is:

$$\nabla A = 6.$$

The Inculcation of ∇abc in terms of the Trichotomous Upright Right Triangle Area can be rewritten to illustrate the equity of the 3-4-5-6 Golden Upright Right Triangle Sides with the Cartesian Coordinates in the following manner:

$$\nabla A = [ab]/2 = [ab] \div 2 = [xy]/2 = [xy] \div 2$$

because

$$[ab]/2 = [xy]/2 \text{ and } [ab] \div 2 = [xy] \div 2$$

Where,

$\nabla a = a = x$; and
$\nabla b = b = y$.

Holistically, the "Total Trichotomous Triune Calculative Characteristics of ∇abc" are expressed in a mathematical area definition as follows:

$$\nabla A \text{ of } \nabla abc = [\text{Side } a \cdot \text{Side } b] \div 2 = [4 \cdot 3] \div 2 = 6,$$

$$\left.\begin{array}{l} a = 4; \\ b = 3; \\ c = 5; \text{ and} \\ A = 6. \end{array}\right\} \;\; \text{\textit{Vabc}}$$

The Geometric Explanation of the "∇ Triangle Interpolation Equation for Side c" using Trioengineering Notation

Illustrating the "ulterior" characteristics of the 3-4-5-6 GURT according to the " ∇Triangle Interpolation Equation of ∇abc" (also known as the geometric " ∇Triangle Interpolation Equation for Side c"):

$$\nabla y = y_1 + (\nabla x - x_1)\frac{(y_2 - y_1)}{(x_2 - x_1)}$$

Where, $(x_1, y_1) = (0, 0)$ for the initial intercept origin point "b", thus, $b = 0$, because, ∇abc rests exactly on the y–axis (the ordinate) and ∇abc is located in the 1st Quadrant (∇positive$_{[x]}$, ∇positive$_{[y]}$) of the Cartesian Coordinates: " I. ", as a true shape that has tangible magnitude (size) and distance as it is an exact part of the "Visualus Isometric Cuboid" (that naturally has magnitude and distance as a tri–coordinate shape and form) the "b" therefore is the "y–intercept" as the origin point for $\nabla abc = (x_0, y_0)$, because Side b is on the y–axis, thus, the " ∇Triangular Slope–Intercept Equation" is: $\nabla[\nabla y = \nabla m \nabla x + \nabla b] = \nabla[\nabla y = \nabla mx + \nabla b]$, defined mathematically and geometrically (using Trioengineering Notation) as:

$$\nabla[y = \nabla mx + \nabla b],$$

Note: $\nabla[y = \nabla mx + 0]$,

Thusly, $\nabla[y = \nabla mx]$, because,

$$[\nabla m] = 1 \text{ for } \quad = \quad = \quad =$$

$$[upr]\ b\left[\ \ \right] b \quad =$$

$$[inv]$$

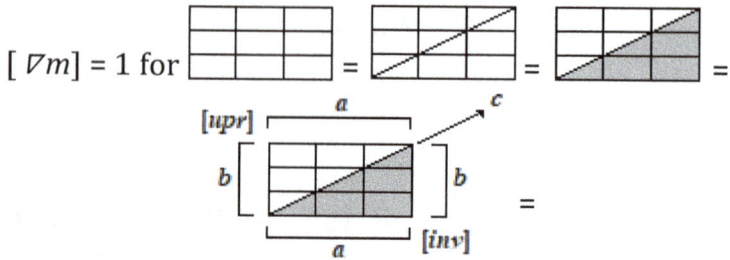

$$\frac{\Delta[\nabla y]}{\Delta[\nabla x]} = \frac{Changes\ in\ Triangular\ Coordinate\ y}{Changes\ in\ Triangular\ Coordinate\ x} = \frac{(y_2 - y_1)}{(x_2 - x_1)} = \frac{(y_2 - 0)}{(x_2 - 0)} = \frac{y_2}{x_2} = \frac{same\ number}{same\ number} = \frac{\nabla rise}{\nabla run}$$

$$= 1,$$

As such $[m = 1]$ can then be substituted yielding the following: $\nabla[y = (1)x]$, thus, $[\nabla y = \nabla x]$ with a slope of "1" (within the confines of the "3 by 3 Standard Table Format"), therefore representing an exact one to one ratio for $x_{0...3}$ to $y_{0...3}$ or a "1:1 ratio" for $x_{0...3}$ exactly matching $y_{0...3}$ for all points that construct "Side c" of "∇abc". Note that "$\nabla b = 0$" as the "∇y–intercept" because Side b rests precisely on the y-axis.

Thusly,

The "Side c ∇Right Trichotomous Triangle Interpolation Equation Interpretations"

The Final "Side c ∇Triangle Interpolation Equation" is written using Trioengineering Notation in the following manner:

Side $c = \nabla y = 0 + (\nabla x - 0)\dfrac{(y_2 - 0)}{(x_2 - 0)} = \nabla y = \nabla x$, as such, $\Delta[\nabla y] = \Delta[\nabla x]$ as a direct 1:1 ratio for .

Chapter Three follows and describes the Concepts of Trichometry.

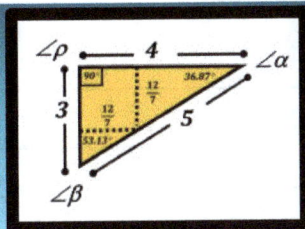

Ask, and it shall be given you.

Matthew 7: 7

Trichometry © Defined

Trichometry pronounced: ["Try" · "Kom" · "Met" · "Tree"]—broadly defined "The study of the geometrics of the Golden Upright Right Triangle in Cartesian Coordinates". A more scientific definition of "Trichometry" is as follows: "Trichometry = The Trichotomous Metrics (i.e., "Measurement") in and as the 3-4-5-6 Golden Upright Right Triangle (also referred to by the acronym "GURT")".

A mathematical model that illustrates the abovementioned in geometric measures is as follows:

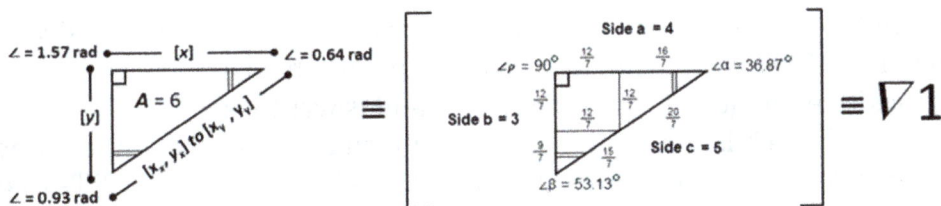

Note the following:

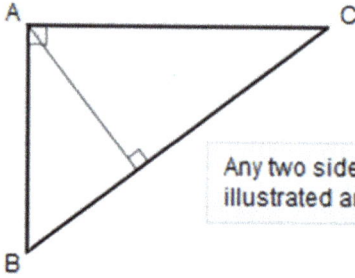

A

C

B

Any two sides of the right triangles illustrated are the Catheti.

Because, A = 6 and P = 12, where, The following is mathematically defined using "Trioengineering Notation" (that uses the "∇" mathematical geometric symbol form Trioengineering to represent the GURT) mathematically define "GURT–TIE" or the "GURT Tripositive Insular Equation" = "$\nabla[P \div A]$" is therefore calculated as $\nabla[P \div A] = \nabla[12 \div 6] = \nabla[2]$ this is then plugged into the equation "Trichotomous Tripositive Equation" or "$\nabla[P \div A] \div 2$" which is represented here as: ["$\nabla[2] \div 2$"] The number 2 is the GURT "Total Tripositive Number of Internal Catheti". The "Trichotomous Tripositive Equation" is complete when an anchor line is drawn from the original 90 right angle needed to complete The "Trichotomous Tripositive Equation". The "Trichotomous Tripositive Equation" is now mathematically defined as the difference between the "GURT Tripositive Insular Equation" and the "Total Tripositive Number of Internal Catheti". The "Trichotomous Tripositive Equation" is now complete and is written as: "$\nabla[2]$ divided by the 2 (which is the total number of internal GURTs that are created by the drawn anchor line and is also the total two internal GURTs that surround the GURT Inscribed Square). Thus, the "Trichotomous Tripositive Equation" is now calculated as:

$$\text{"} \nabla[P \div A] \div 2\text{"} = \text{"} \nabla[12 \div 6] \div 2\text{"} = \nabla[2] \div 2 = \text{"} \nabla 1\text{"}.$$

The Three Trichotomous In-Depth Operations of Mathematics of Trichometry

The mathematics of Trichometry is composed of three trichotomous triple independent in-depth operations that are:

1.) **"Intercalation"** that is symbolized and written as an acronym in the following "Tripositive" format: **"[int]"** and is geometrically mathematically defined as follows—The insertion of internal calculations, measurements, and operations inherent to the 3-4-5-6 GURT;

2.) **"Interpolation"** that is symbolized and written as an acronym in the following "Tripositive" format: **"[inp]"** and is geometrically mathematically defined as follows—The introduction of additional mathematical geometric terminology between and within 3-4-5-6 GURT definitive mathematical operations; and

3.) **"Intercalculation"** that is symbolized and written as an acronym in the following "Tripositive" format: **"[inc]"** and is geometrically mathematically defined as follows—The term "Intercalculation" is a portmanteau of the prefix "Inter" and the term "Calculation" that literally means to internally determine through in-depth mathematical operations the entirety of the 3-4-5-6 GURT.

Graphical Geometric Representations of the GURT

In terms of external area geometrics, the GURT has the following characteristics (that are illustrated here graphically for ease of comprehension):

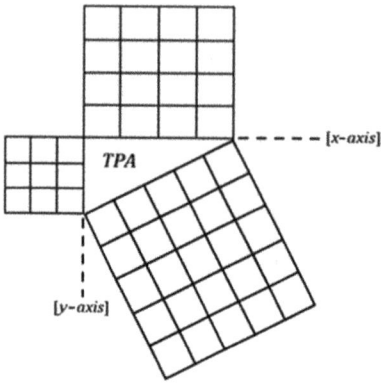

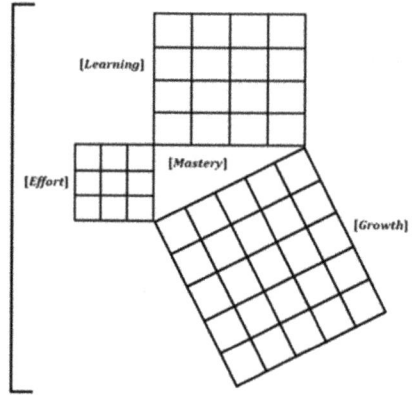

=

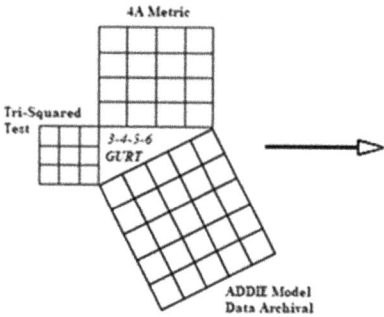

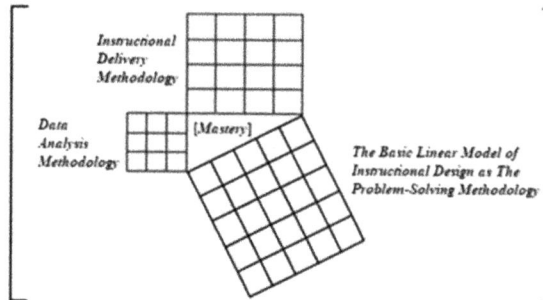

=

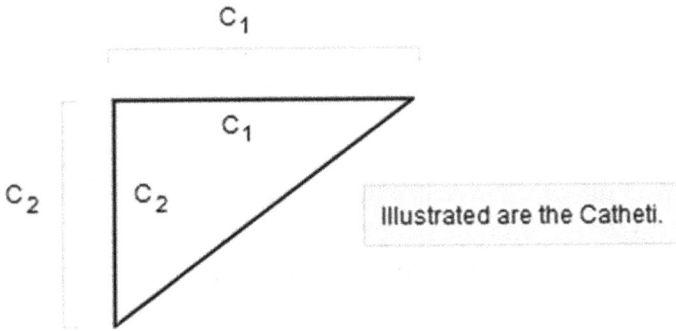

C_1

C_1

C_2 C_2

Illustrated are the Catheti.

The Three Trichotomous Laws of Trichometry

The **"Three Trichotomous Laws of Trichometry"** are the primary rules that govern Trichometry as a mathematical discipline. They are:

Law 1.

Trichometry only deals with all of the aspects of the "3-4-5-6 Golden Upright Triangle" also known by the acronym "GURT".

Law 2.

There are **no negatives** as there exist **"no negative measures in geometrical scalar space"**. This is especially true of Trichometry as it is a Triological Science that has all of its measures only in the 1st Cartesian Coordinate where all coordinates are: (positive, positive).

Law 3.

The true measures as the "Extent" of the GURT have the specific measurements of "1-2-3-4-5-6" that are equal to the following: (1.) $\angle \alpha$ = 36.87°; (2.) $\llcorner \rho$ = 90°; (3.) $\angle \beta$ = 53.13°; (4.) Side a = 4; (5.) Side b = 3; (6.) Side c = 5; and an overall Area ("A") = 6. The shortened "3-4-5-6 GURT" nomenclature is used for the specific purposes of parsimony.

Trichotomous Scalar Numbers

The "Trichotomous Scalar Number" is defined in terms of the mathematics of Trichometry as follows: "Trichotomous Scalar Numbers are positive integers located only in the first Cartesian Coordinate = Quadrant One. Trichotomous Scalars are also a value of a continuous quantity that can represent a distance along a line that is used in the mathematics of "linear algebra" the branch of mathematics that is concerned with linear equations of the form:

$$\text{``}(a_1 x_1 + \ldots + a_n x_n = b)\text{''}.$$

Trichotomous Scalar Measures

The following table provides a trichotomous list of "Trichotomous Scalar Numbers" as "Trichotomous Scalar Measures" using Triological Scientific "Trioengineering Notation". The three trichotomous scalar measures are identified as the: *"Trichotomous Scalar Versor"* (*" Ʋversor"* = The *"Trioengineered Versor"*); *"Trichotomous Scalar Vector"* (*" Ʋvector"* = The *"Trioengineered Vector"*); and the *"Trichotomous Scalar Tensor"* (*" Ʋtensor"* = The *"Trioengineered Tensor"*).

Trichotomous Scalar Versor	Trichotomous Scalar Vector	Trichotomous Scalar Tensor
Written as: "*Vversor*"	Written as: "*Vvector*"	Written as: "*Vtensor*"
Name:	Name:	Name:
"*Trioengineered Versor*"	"*Trioengineered Vector*"	"*Trioengineered Tensor*"
Defined as:	Defined as:	Defined as:
A generalization of two-coordinate numbers to three coordinates originally devised to measure mechanics problems with a "Trichotomous Norm" = "∇ 1" as a "Trichotomous Mathematical Function" = A "Trichotomous Expression", which is a "Trichotomously Defined Rule" that follows the "Three Trichotomous Laws of Trichometry" and that illustrates the unique holistic relationship between the sides of the 3-4-5-6 Golden Upright Right Triangle (i.e., the "GURT"). The "Trioengineering Model" =	A tangible trichotomous linear unit with given unit measurable size of one 3-4-5-6 GURT that travels along the given path of the entirety of the 3-4-5-6 Golden Upright Right Triangle and returns to the origin point that is the first angle ("*angle alpha*" = $\angle$ α = 36.87 degrees (as "36.87°" = $[x_x, y_y]$ = (4, 3) as the Cartesian Coordinates. The model that follows illustrates how the Trioengineered Vector heads in a path in a specified trajectory that follows in a counter-clockwise pattern. "Trioengineering Vector Model":	A "Trioengineered Tensor" is an in-depth quantitative and qualitative "Trichotomous Mathematical Object" that models the physical relationship of the given "Trichotomous Quantities" using Triological Science of "Ternary Algebra". The ideal model of the "Trioengineered Tensor" is represented by the mathematical model of the 3-4-5-6 Golden Upright Right Triangle as follows (illustrating the multilinear relationship between the sets of GURT trichotomous Ternary Algebra objects in tricoordinate trichotomous vector-defined space):
. As a prime example of a "Trichotomous Scalar Versor" that is a "Trioengineered Versor" ("*Vversor*").	. As a prime example of a "Trichotomous Scalar Vector" that is a "Trioengineered Vector" ("*Vvector*").	. As a prime example of a "Trichotomous Scalar Tensor" that is a "Trioengineered Tensor" ("*Vtensor*").

The Trichotomous Scalar Graphic Model

The Trichotomous Scalar Graphic Model illustrates the 3-4-5-6 GURT in the 1st Quadrant of Cartesian Coordinates that is (positive, positive).

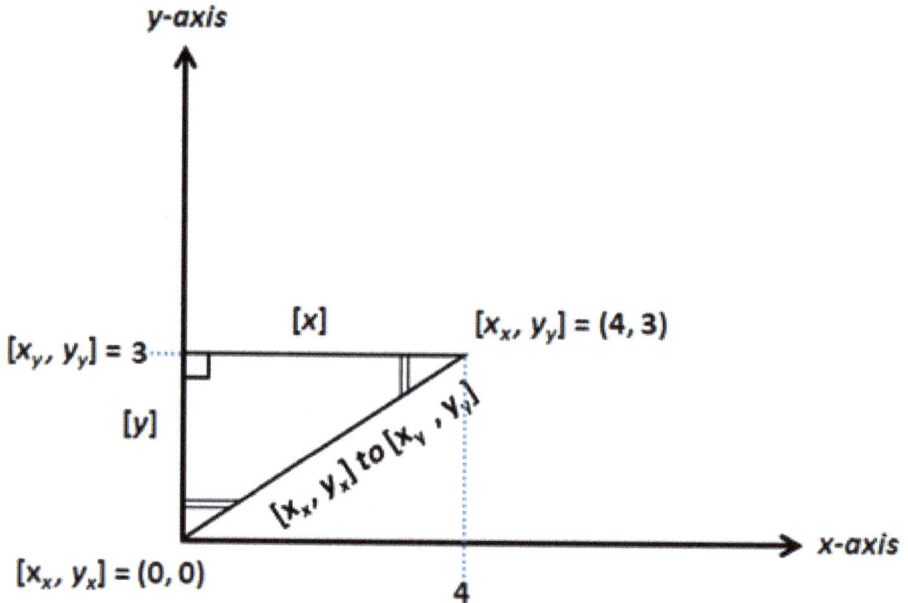

Explaining the External Characteristics of the GURT as the "Extent"

The "Extent" of the mathematics of Trichometry as a mathematical definition: "The Extent of the GURT is used to explain the characteristics of the 3-4-5-6 Golden Upright Right Triangle and uses the acronym "ex" as a "Tripositive" (that uses the square brackets "[ ]" to specifically concentrate and focus on the internal trichotomous parameters placed inside of the brackets) in explicative mathematical operation called: "**Trioengineering Notation**". This notation "expresses" the "external" characteristics that are the "extent" in the written form—" $\sqrt{}$[ex]" as an external GURT mathematical explicative."

Examples of the Trichometry Extent

" $\nabla[ex]_{[a]} \equiv 4$ ", literally means—"The Trioengineered Trichotomous Tripositive extent of Tripositive Side a is identical to 4" and is mathematically defined as: "The Trioengineered 3-4-5-6 GURT Triangular Model Tripositive extent of Tripositive Side a is identical to (and is the trichotomous: extent of; the external; expression to) 4."

This is further explained by the graphical model that follows below.

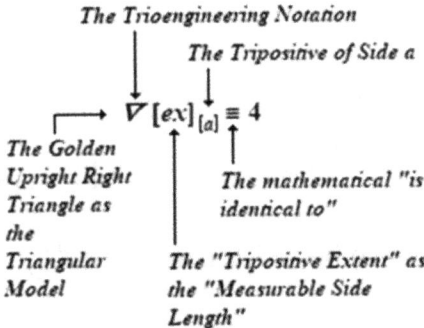

The Trioengineering Notation

The Tripositive of Side a

$$\nabla[ex]_{[a]} \equiv 4$$

The Golden Upright Right Triangle as the Triangular Model

The mathematical "is identical to"

The "Tripositive Extent" as the "Measurable Side Length"

Further Trichometry Extent Measures

Examine the following Model:

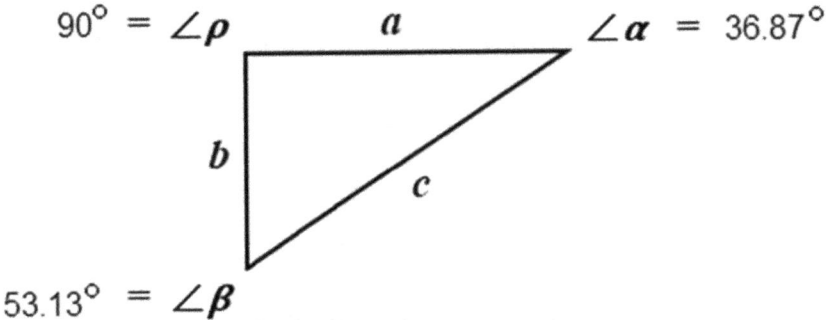

$90° = \angle\rho$ a $\angle\alpha = 36.87°$

b

c

$53.13° = \angle\beta$

Side a = ◸∠***α*** to ◸∟***ρ*** = Side ***αρ*** or "***αρ***" = 4, Thus, a = "***αρ***" or "Trioengineered angle alpha to Trioengineered angle rho";

Side b = ◸∟***ρ*** to ◸∠***β*** = Side ***ρβ*** or "***ρβ***" = 3, Thus, a = "***ρβ***" or "Trioengineered angle rho to Trioengineered angle beta"; and

Side c = ◸∠***β*** to ◸∠***α*** = Side ***βα*** or "***βα***" = 5, Thus, a = "***βα***" or "Trioengineered angle beta to Trioengineered angle alpha".

As such,

Extent = "***ex***";

"***ex***" ≡ "$\overset{\sqcup}{ex}$" = "The External Extent of the 3-4-5-6 GURT Measure";

◸$[ex]_{[a]}$ ≡ 4 and can also be written as: "**◸$[\overset{\sqcup}{ex}]_{[a]}$ ≡ 4**";

◸$[ex]_{[b]}$ ≡ 3 and can also be written as: "**◸$[\overset{\sqcup}{ex}]_{[b]}$ ≡ 3**"; and

◸$[ex]_{[c]}$ ≡ 5 and can also be written as: "**◸$[\overset{\sqcup}{ex}]_{[c]}$ ≡ 5**".

Chapter Four follows and describes Trichometry Geometric Concepts.

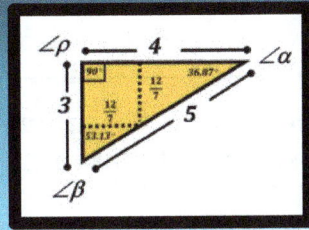

When I sit in darkness, the Lord shall be a light unto me.

Micah 7: 8

Trichometry Unit Measurement in Terms of Geometry

The Geometric "Sine" = The Tripositive Trichometry Calculation = [Side b ÷ Side c] = [3 ÷ 5] = [0.6];

The Geometric "Cosine" = The Tripositive Trichometry Calculation = [Side a ÷ Side c] = [4 ÷ 5] = [0.8];

The Geometric "Sine" Squared = The Tripositive Trichometry Calculation Squared = ["Sine"]^2 = [3 ÷ 5]^2 = [0.6]^2 = 0.36; and

The Geometric "Cosine" Squared = The Tripositive Trichometry Calculation Squared = ["Cosine"]^2 = [4 ÷ 5]^2 = [0.8]^2 = 0.64.

$\triangledown[0.6]^2 + \triangledown[0.8]^2 = \triangledown 1$; is equal to the following,

$\triangledown[0.36] + \triangledown[0.64] = \triangledown 1$;

$\triangledown[\angle]^2 + \triangledown[\diagup]^2 = \triangledown 1$; thus,

$\triangledown[3 \div 5]^2 + \triangledown[4 \div 5]^2 = \triangledown 1$.

113

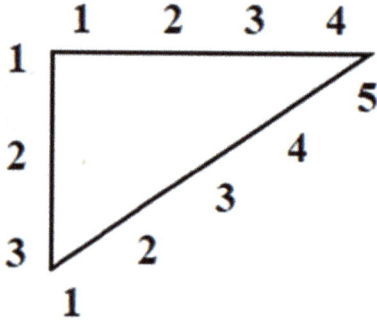

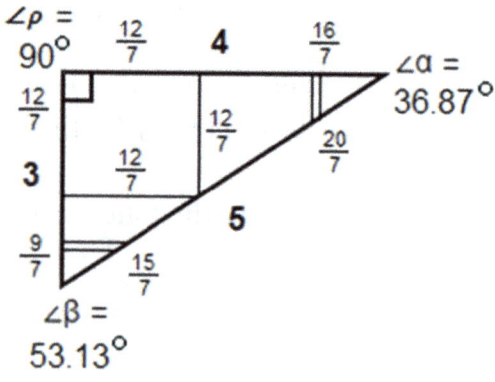

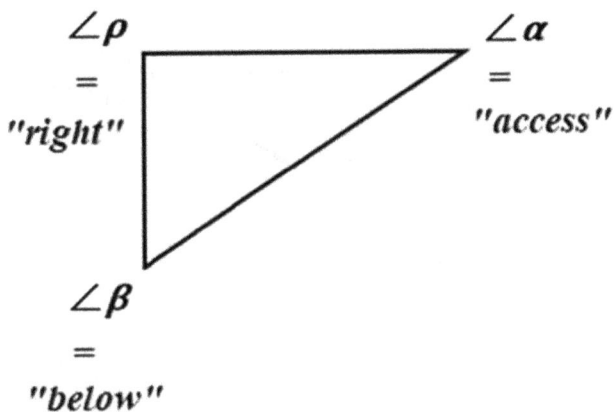

$\angle\rho$
=
"right"

$\angle\alpha$
=
"access"

$\angle\beta$
=
"below"

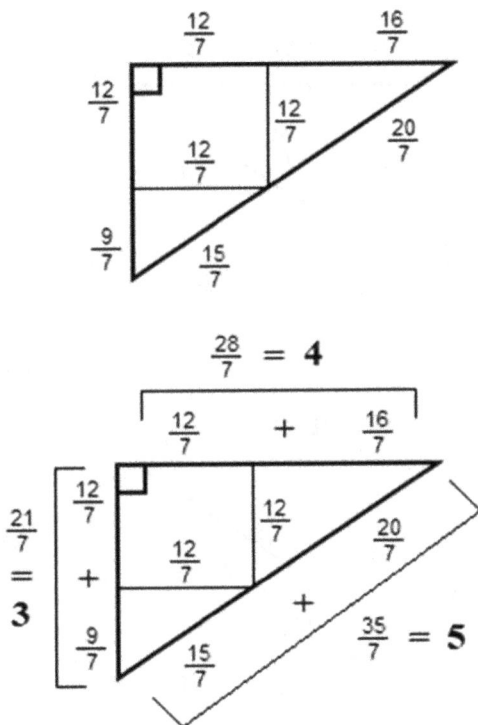

TRICHOMETRY ™ © *The Study of the Geometrics of the 3-4-5-6 Golden Upright Right Triangle in Cartesian Coordinates.* Osler Studios Incorporated ™ © Copyright 2022 All Rights Reserved.

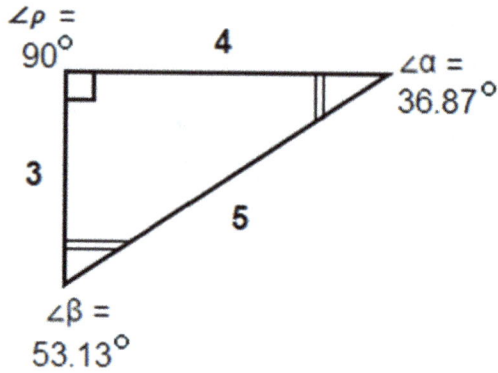

∠ρ = 90°   4   ∠α = 36.87°

3

5

∠β = 53.13°

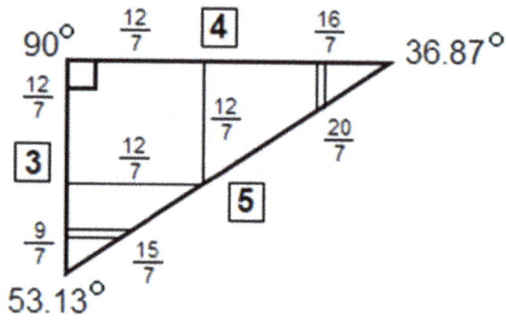

90° $\frac{12}{7}$ 4 $\frac{16}{7}$ 36.87°

$\frac{12}{7}$

$\frac{12}{7}$

3    $\frac{12}{7}$ $\frac{20}{7}$

5

$\frac{9}{7}$ $\frac{15}{7}$

53.13°

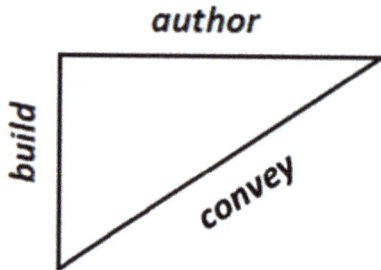

author

build

convey

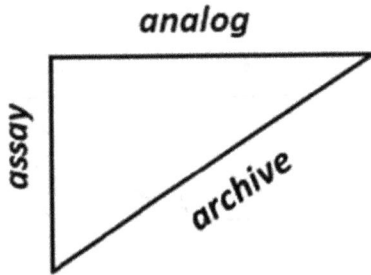

analog

assay

archive

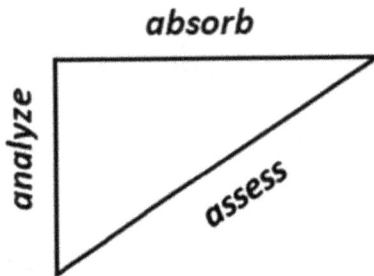

absorb

analyze

assess

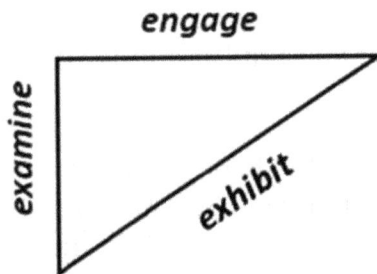

engage

examine

exhibit

Additional Trichometry Properties of the 3-4-5-6 Golden Upright Right Triangle

The graphical model for the external "extent" of the 3-4-5-6 GURT based upon the internal "Inscribed Square" of the GURT that is graphically represented in the graphical model that follows.

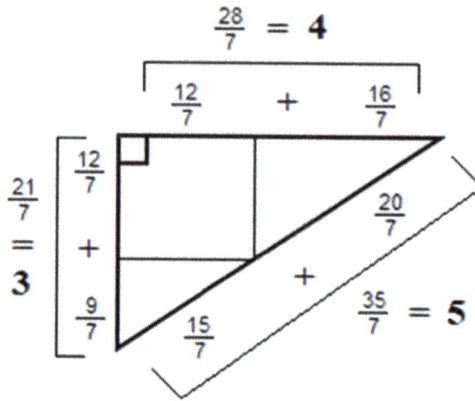

$$\frac{28}{7} = 4$$

$$\frac{12}{7} \quad + \quad \frac{16}{7}$$

$\frac{12}{7}$

$\frac{21}{7} = 3$

$= + 3$

$\frac{9}{7}$

$\frac{20}{7}$

$\frac{15}{7}$

$+$

$\frac{35}{7} = 5$

Numerical Properties and their respective definitions:

1 = Representing the 1 Whole Upright Right Triangle (and also represents the 1 internal Inscribed Square);

2 = Representing the 2 Inner Right Triangles Created from the Right-Angle Anchor Line (and also represents the 2 Internal Upright Right Triangles that enclose the Internal GURT Inscribed Square);

3 = Representing the 3 Sides (and also the 3 Angles) for the Entire Large Upright Right Triangle;

4 = Representing the 4 GURT Shapes that includes—the 3 Upright Right Triangles (that are the Entire GURT and the 2 Internal Upright Right Triangles that enclose the Internal GURT Inscribed Square) + the 1 Inscribed Square;

5 = Representing the Unit Length of the "Trichometry Inclination" = **"Trichometric Incathetus"** (a term unique to the mathematics of Trichometry) that is equal to the Geometric "hypotenuse" that is also the Unit Length of the 3-4-5-6 GURT "[Side *c*]" thus, the "Trichometry Inclination" = **"Trichometric Incathetus"** = "hypotenuse" = the 3-4-5-6 GURT "Side *c*";

6 = Representing the 6 Internal Angles for the Internal Inscribed Square Shape within the 3-4-5-6 GURT that is composed of the six 90° Right Angles including the "rho" = $\angle\rho$ (and the other three 90° Inscribed Square Angles along with the two External 90° Angles on the outside of the Inscribed Square), (there are also six Sides for two Internal Upright Right Triangles created by the Inscribed Square);

7 = Representing the 7 the "Unique 3-4-5-6 GURT Constant Unit Denominator" for all GURT subsection side lengths based upon the internal GURT Inscribed Square;

8 = Representing the 8 Unique 3-4-5-6 GURT Side Length Subsections created by the GURT Inscribed Square (this includes the two internal GURT Inscribed Square Side Lengths to create the grand total of eight); and lastly,

9 = Representing the 9 Sides of all three Upright Right Triangles (composed of the entire GURT and the two Internal Upright Right Triangles created by the internal Inscribed Square, for a grand total of nine sides).

A More In-Depth Segmentation of the Internal 3-4-5-6 GURT Characteristics

The following series of graphical models are designed to aid in the further comprehension of the internal and external GURT characteristics.

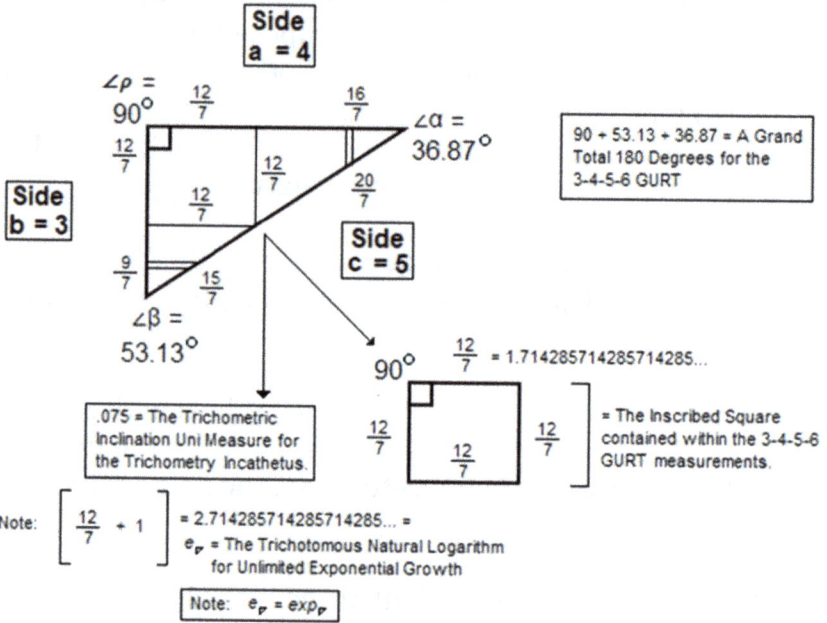

Side
a = 4

∠ρ =
90°

$\frac{12}{7}$

$\frac{16}{7}$

∠α =
36.87°

$\frac{12}{7}$

Side
b = 3

$\frac{12}{7}$

$\frac{12}{7}$

$\frac{12}{7}$

$\frac{20}{7}$

Side
c = 5

$\frac{9}{7}$

$\frac{15}{7}$

∠β =
53.13°

90°

$\frac{12}{7}$ = 1.714285714285714285...

90 + 53.13 + 36.87 = A Grand
Total 180 Degrees for the
3-4-5-6 GURT

.075 = The Trichometric
Inclination Uni Measure for
the Trichometry Incathetus.

$\frac{12}{7}$

$\frac{12}{7}$

$\frac{12}{7}$

$\frac{12}{7}$

= The Inscribed Square
contained within the 3-4-5-6
GURT measurements.

Note: $\left[\frac{12}{7} + 1 \right]$ = 2.714285714285714285... =
e_p = The Trichotomous Natural Logarithm
for Unlimited Exponential Growth

Note: $e_p = \exp_p$

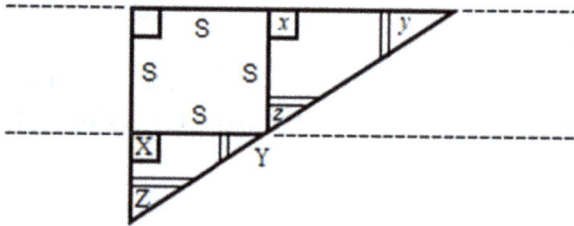

The following applies to geometrically define the last two GURT graphical models to determine precisely how the GURT 12/7 Inscribed Square side length is determined:

Side $a = 4$; Side $b = 3$; and finally Side $c = 5$, thus, X = x; Y = y; and Z = z, therefore, Side XY = S; Side XZ = $3 - s$; Side $xy = 4 - S$; and Side $xz = S$. As such, $\frac{Side\ XY}{Side\ XZ} = \frac{Side\ xy}{Side\ xz}$. This then leads to the following: $\frac{S}{3-S} = \frac{4-S}{S}$, now solve the equation by cross multiplying which provides the following answer:

$$(S)(S) = (3 - S)(4 - S)$$

$$S^2 = 12 - 7S + S^2$$

Next,

$$S^2 - \left(-7S + S^2\right) = 12 - 7S + S^2 - \left(-7S + S^2\right)$$

$$\cancel{S^2} + 7S \cancel{-S^2} = 12 \cancel{-7S + S^2} \cancel{-\left(-7S + S^2\right)}$$

$$7S = 12$$

Divide both sides by 7, which then provides the answer:

$$S = \frac{12}{7}$$

Which is the 3-4-5-6 GURT Inscribed Square side length that is used to create all of the side lengths of the GURT. Note:

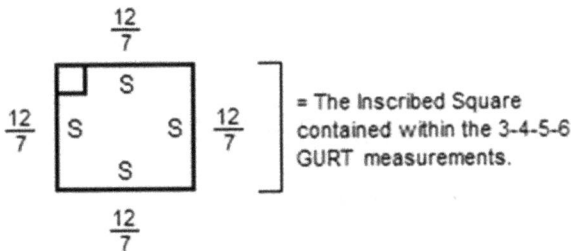

= The Inscribed Square contained within the 3-4-5-6 GURT measurements.

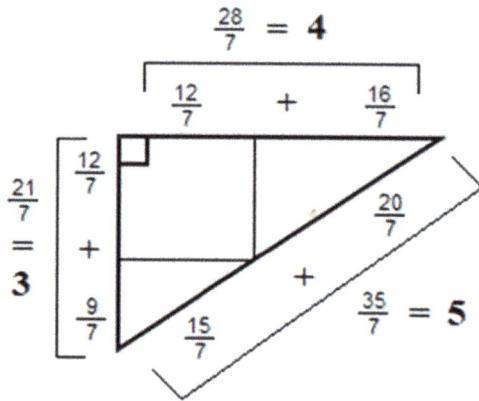

$$\frac{28}{7} = 4$$

$$\frac{12}{7} \quad + \quad \frac{16}{7}$$

$$\frac{21}{7} = 3 \quad \frac{12}{7} + \frac{9}{7}$$

$$\frac{20}{7}$$

$$\frac{15}{7}$$

$$\frac{35}{7} = 5$$

Because, the 3-4-5-6 GURT three Sides are calculated using the GURT Inscribed Square side lengths to determine overall GURT side lengths as follows:

To calculate the full measurements for "Side a" = 4 – 12/7, because, [4 – S] is the side length provided earlier for "Side a" (Note: "Side a" = 4 and "Side xy" = 4 – S), therefore, due to the GURT Inscribed Square side (represented as "S") that appears as a part of "Side a" was previously determined to equal "12/7". Thus, all that is left is to determine the exact total of the remaining part of "Side a" that should with 12/7 equal the "Side a" overall unit measurement of 4. Using the following calculation:

$$4 - S =$$

$$4 - 12/7 =$$

$$4/1 - 12/7 =$$

$$4(7)/1(7) - 12/7 =$$

$$28/7 - 12/7 =$$

$$\underline{16/7}.$$

As such, the following can now be determined:

$12/7 + 16/7 = 28/7 = \underline{\textbf{4}}$ which is the overall 3-4-5-6 GURT Unit measurement side length for "Side a".

To calculate the full measurements for "Side b" = 3 – 12/7, because, [3 – S] is the side length provided earlier for "Side b" (Note: "Side b" = 3 and "Side XZ" = 3 – S), therefore, due to the GURT Inscribed Square side (represented as "S") that appears as a part of "Side b" was previously determined to equal "12/7". Thus, all that is left is to determine the exact total of the remaining part of "Side b" that should with 12/7 equal the "Side b" overall unit measurement of 3. Using the following calculation:

$$3 - S =$$

$$3 - 12/7 =$$

$$3/1 - 12/7 =$$

$$3(7)/1(7) - 12/7 =$$

$$21/7 - 12/7 =$$

$$\underline{9/7}.$$

As such, the following can now be determined:

$9/7 + 12/7 = 21/7 = \underline{\textbf{3}}$ which is the overall 3-4-5-6 GURT Unit measurement side length for "Side b".

To calculate the full measurements for Side c it is necessary to determine the unit measurement of the inclination side which is the unit length of the "Trichometry Inclination" = **"Trichometric Incathetus"** that is equal to the Geometric "hypotenuse" that is also the Unit Length of the 3-4-5-6 GURT "Side c" thus, the "Trichometry Inclination" = "Trichometric Incathetus" = "hypotenuse" = the 3-4-5-6 GURT "Side c" as previously written and is calculated as follows using the Pythagorean Theorem:

$$c^2 = a^2 + b^2$$

Therefore,

$$\sqrt{c^2} = \sqrt{a^2 + b^2}$$

Thus,

$$c = \sqrt{a^2 + b^2}$$

Relating to "Side c" the Inscribed Square lower Upright Right Triangle below the Inscribed Square determined the following measurements previously that $S = 12/7$, "Side a" = $12/7$ (which is the base of the GURT Inscribed Square), and "Side b" = $9/7$. "Side c" can now be calculated by using the mathematical equation for "Side c" that was derived from the Pythagorean Theorem above = "$c = \sqrt{a^2 + b^2}$". The calculation for "Side c" is as follows:

$$c = \sqrt{a^2 + b^2} =$$

$$c = \sqrt{\left(\frac{9}{7}\right)^2 + \left(\frac{12}{7}\right)^2} =$$

$$\sqrt{\frac{81}{49} + \frac{144}{49}} =$$

124

$$\sqrt{\frac{225}{49}} =$$

$$\frac{\sqrt{225}}{\sqrt{49}}$$

<u>15/7</u>.

To calculate the full measurements for GURT "Side c" can now be calculated. It was previously determined that a portion of the overall 3-4-5-6 GURT the smaller Upright Right Triangle calculation of its "Side c" = 15/7, therefore, the full side length of the overall GURT "Side c" has the following equation: [5 – 15/7]. This is then calculated by using the following equation:

$$5 - 15/7 =$$

$$5/1 - 15/7 =$$

$$5(7)/1(7) - 15/7 =$$

$$35/7 - 15/7 =$$

<u>20/7</u>.

As such, the following can now be determined:

15/7 + 20/7 = 35/7 = **5** which is the overall 3-4-5-6 GURT Unit measurement side length for "Side c".

The 3-4-5-6 Golden Upright Right Triangle is equal in degrees to one half of a circle also known as a "semicircle". The graphical models that follow illustrate the similarities and equality of the two shapes that each have a grand total of 180°.

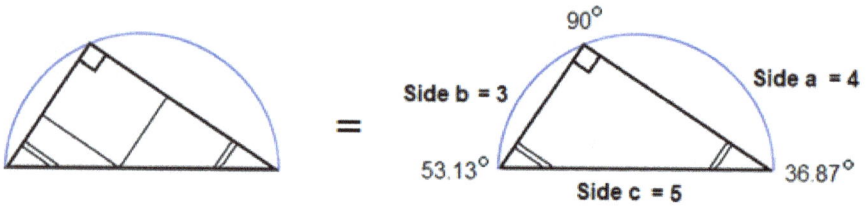

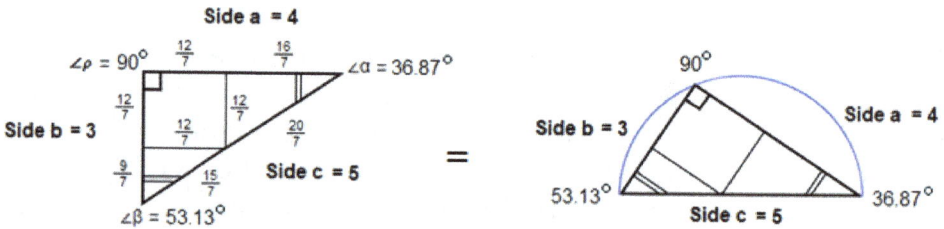

Thales' Theorem particularly applies to semicircles and a right triangle inscribed within a semicircle. Thales' Theorem states the following as it directly relates to the 3-4-5-6 GURT: "That if a given line is determined to be at angles "alpha" (36.87°) and "beta" (53.13°) respectively and is then established as the diameter of a given circle and angle "rho" (90°) is established as a given point on the remaining semicircle's (i.e., "half-circle's") circumference (in this case), then the angle that touches the semicircle's circumference is determined to be a right angle." A Semicircle that contains the 3-4-5-6 Golden Upright Right Triangle is called the "Golden Semicircle".

Chapter Five follows and examines Trichometry Operations.

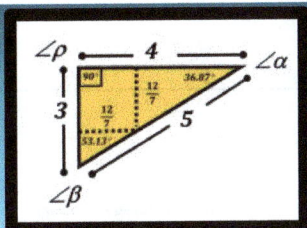

Thou wilt show me the path of life.

Psalm 16: 11

Further Trichometric Proportions of the 3-4-5-6 Golden Upright Right Triangle expressed in Graphical Models

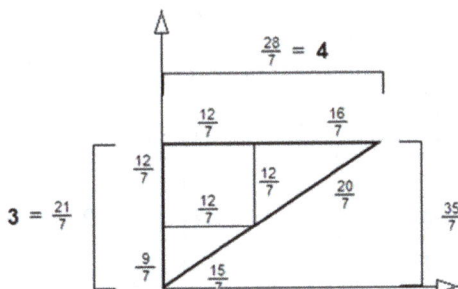

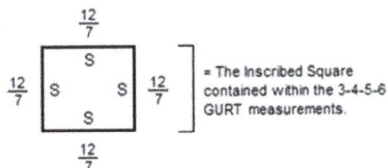

= The Inscribed Square contained within the 3-4-5-6 GURT measurements.

Using Trioengineering Notation:

$\nabla y = \nabla mx + \nabla b;$

$\nabla y = \nabla \frac{3}{4}x + \nabla b;$ and

$\nabla y = \nabla \frac{3}{4}x + \nabla 3;$

Where,

$\nabla[y = \frac{3}{4}x + 3],$ $\nabla[\text{Side } a = 4]$ and $\nabla[\text{Side } b = 3];$ and

Therefore the following is true:

$\nabla[y = \frac{3}{4}(0) + 3]$ and yields $\nabla y = \nabla 3.$

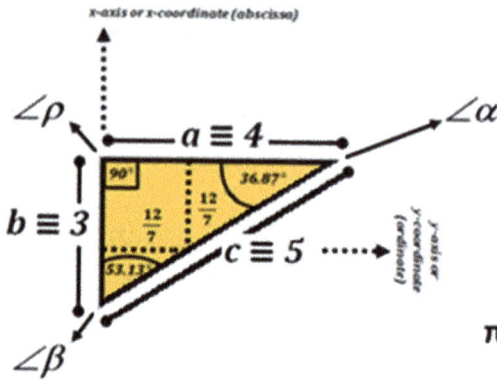

$$\varphi_{\nu} \equiv \dfrac{1+\sqrt{3\sqrt{1+\left[\frac{4}{3}\right]^2}}}{2}$$

x-axis or x-coordinate (abscissa)

∠ρ

$a \equiv 4$

∠α

$90°$ $36.87°$

$b \equiv 3$ $\frac{12}{7}$ $\frac{12}{7}$

$53.13°$ $c \equiv 5$

∠β

The Inclination of $c \equiv \left[\frac{3}{4}\right] \equiv 0.75 \equiv [$"╱"$]$

$$c \equiv \left[b\sqrt{1+\left[\frac{4}{3}\right]^2}\right] \qquad e_{\nu} \equiv \left[\frac{12}{7}+1\right]$$

$$\pi_{\nu} \equiv \left[\dfrac{4}{\sqrt{\dfrac{1+\sqrt{3\sqrt{1+\left[\frac{4}{3}\right]^2}}}{2}}}\right] \equiv$$

$$\pi_{\nu} \equiv \left[\dfrac{a}{\sqrt{\varphi_{\nu}}}\right] \equiv \pi_{\nu} \equiv \left[\dfrac{4}{\sqrt{\varphi_{\nu}}}\right]$$

The Graphical Models above illustrates the "Trichotomous Golden Ratio"; the "Trichotomous Golden Pi"; and the "Trichotomous Golden Growth Exponential" respectively. Further Trichometric Proportions of the 3-4-5-6 Golden Upright Right Triangle expressed in additional graphical models follow.

128

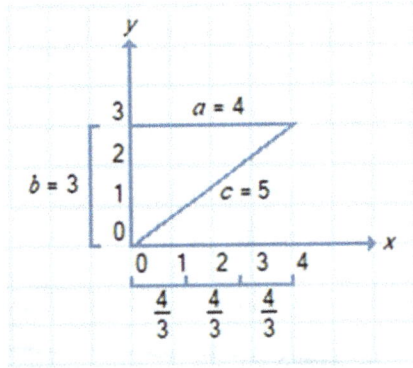

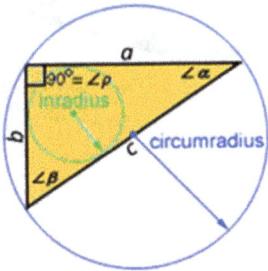 ≡ 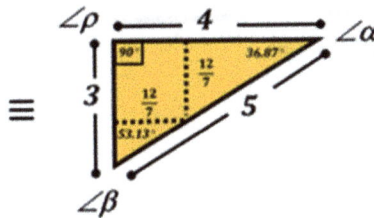 ≡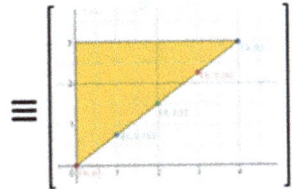

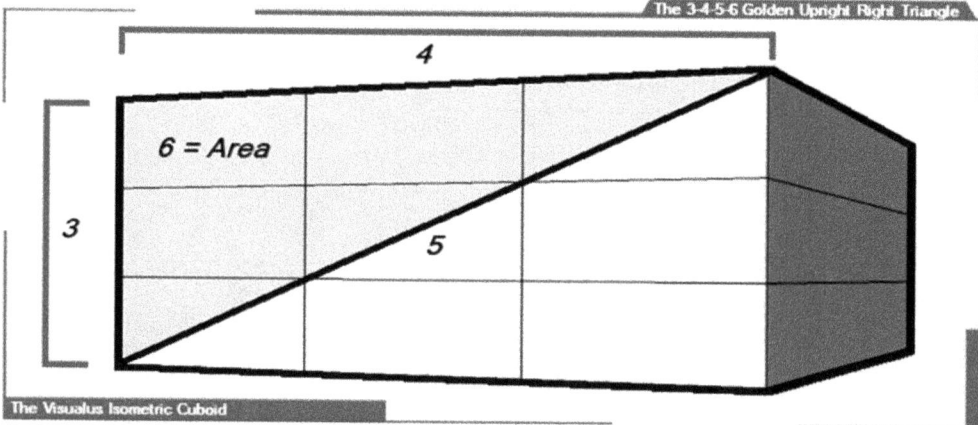

4

6 = Area

3

5

The Visualus Isometric Cuboid

According to the "Triangular Interpolation Equation of ∇abc"
(also known as the "Side c: ∇ Triangular Interpolation Equation"):

$\nabla y = y_1 + (\nabla x - x_1)\dfrac{(y_2 - y_1)}{(x_2 - x_1)}$; where, $(x_1, y_1) = (0, 0)$ for the initial intercept origin point "b", thus, $b = 0$, because,

∇abc rests exactly on the $y - axis$ (the ordinate) and ∇abc is located in the 1st Quadrant (positive, positive) of the
Cartesian Coordinates as a true shape that has tangible magnitude (size) and distance as it is an exact part of the
"Visualus Isometric Cuboid" (that naturally has magnitude and distance as a tri – coordinate shape and form) the
"b" therefore is the "y – intercept" as the origin point for $\nabla abc = (x_0, y_0)$ because Side b is on the y – axis, thus,
the "∇ Triangular Slope – Intercept Equation" is $\nabla [y = mx + b]$, note: $\nabla [y = mx + 0]$, thus, $\nabla [y = mx]$, because,

$[m] = 1 = \dfrac{\Delta[\nabla y]}{\Delta[\nabla x]} = \dfrac{\text{Changes in Triangular Coordinate } y}{\text{Changes in Triangular Coordinate } x} = \dfrac{(y_2 - y_1)}{(x_2 - x_1)} = \dfrac{(y_2 - 0)}{(x_2 - 0)} = \dfrac{y_2}{x_2} = \dfrac{\text{same number}}{\text{same number}} = \dfrac{\nabla rise}{\nabla run} = 1,$

as such, $\nabla [y = (1)x]$, thus, $[y = x]$ with a slope of "1", representing an exact one to one ratio for $x_{0...3}$ to $y_{0...3}$ or a
$1:1$ for $x_{0...3}$ exactly matching $y_{0...3}$ for all points that construct "Side c" of "∇abc".
Thusly,

Side $c = \nabla y = 0 + (\nabla x - 0)\dfrac{(y_2 - 0)}{(x_2 - 0)} = \nabla y = \nabla x$, as such, $\Delta[\nabla y] = \Delta[\nabla x]$; and there are according to the

aforementioned "Side c: ∇ Triangular Interpolation Equation" that there are trichotomously three separate
"interpretations" of the points that construct Side c that are respectively expressed as follows:
Interpretation 1.
A grand total of four points for ∇abc according to the
For "b" as the "y – intercept" as the origin point for $\nabla abc = (x_0, y_0)$ = when, $x = 0$ then, $y = 0$; as a result the
points that immediately follow are written as:
(x_1, y_1) = when, $x = 1$ then, $y = 1$; (x_2, y_2) = when, $x = 2$ then, $y = 2$; and (x_3, y_3) = when, $x = 3$ then, $y = 3$.
Interpretation 2.
There are only three trichotomous positive points on the positive "incline" and "acclivity" of "side c" that are
represented in the following manner $[(x_0, y_0)$ is not included as it is neutral and not a positive integer]:
(x_1, y_1) = when, $x = 1$ then, $y = 1$; (x_2, y_2) = when, $x = 2$ then, $y = 2$; and (x_3, y_3) = when, $x = 3$ then, $y = 3$.
Interpretation 3.
Lastly, there are two points in the "∇ Triangular Interpolation Equation of Side c" (excluding the end points) that
are located inbetween the end points of (x_0, y_0) and (x_3, y_3) respectively as a true interpolation of Side c:
(x_1, y_1) = when, $x = 1$ then, $y = 1$; and (x_2, y_2) = when, $x = 2$.

Geometric Graphical Models that apply to Trichometry

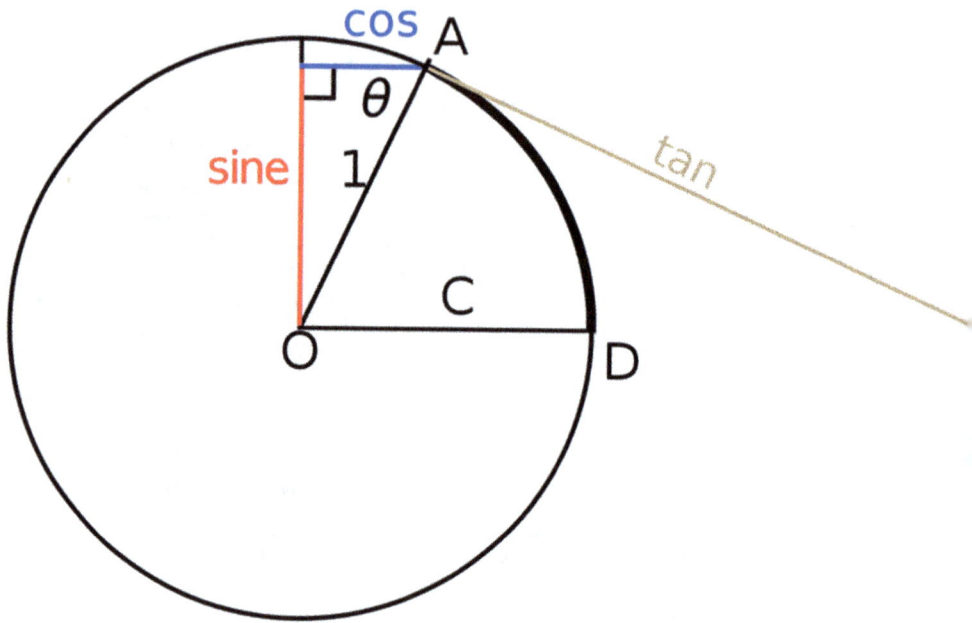

$$\text{Incline: } m = \frac{\Delta y}{\Delta x} = \tan(\theta)$$

Incline: $m = \left[\dfrac{\Delta y}{\Delta x}\right]$ where, $\left[\begin{array}{l} = 3/4 = 0.75 = \textit{to the inclination or} \\ \textit{the diagonal of the triangle, because } \Delta x \\ = \textit{Side a = 4 and } \Delta y = \textit{Side b = 3.} \end{array}\right]$

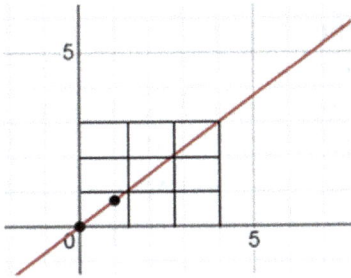

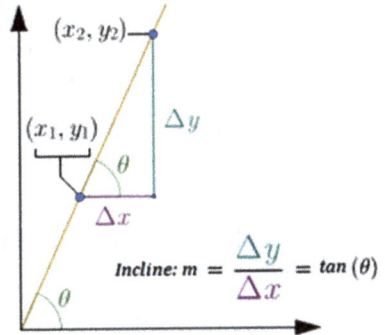

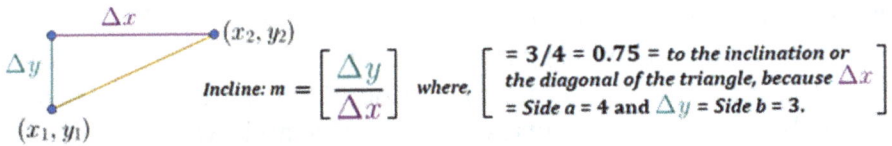

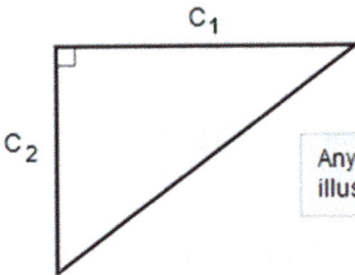

Any two sides of the right triangle illustrated are its Catheti.

The true trichotomous "$e_{\triangleright}$" in terms of "$[\triangleright]$" or the "Golden Upright Right Triangle" with Tricoordinates that are: "Side a" = 4; "Side b" = 3; and "Side c" = 5, respectively, with an "Inscribed Square" with 12/7 = 1.714285714285714285… side lengths. The overall 3-4-5-6 GURT has an inclination = 0.75 as the angled diagonal line = "Side c" = 5. As such, a single 3-4-5-6 GURT Inscribed Square side length + 1 = 2.714285714285714285… which is equal to "$e_{\triangleright}$" = "$exp_{\triangleright}$" as the "Trichotomous Golden Growth Exponential".

The Intra-Trichotomy of Trichometry

An "Intra-Trichotomy" is defined as the external, internal, and frontal characteristics that are a part of the of the whole of the 3-4-5-6 Golden Upright Right Triangle that aid in defining its innate characteristics that can be readily and easily observed (especially when illustrated in a specifically detailed visual graphical model). The "Intra-Trichotomy" of the mathematics of Trichometry can be observed in the section that follows that presents in a series of distinctive and explicative graphical models, the mathematical definitions and the mathematical operations that are the unique characteristics that work together to formulate the unique GURT "Intra-Trichotomy" that as an indelible part of the GURT is illustrated to display the unique mathematics inherent to Trichometry.

The Mathematical Operations that are the "Intra-Trichotomy" that is Unique to the Mathematics of Trichometry

In the mathematics of Trichometry three primary trichotomous mathematical operators are used to measure the 3-4-5-6 GURT. Square brackets "[ ]" are used in Trichometry to denote and indicate the three main and major trichotomous areas that are measurable and apply to the 3-4-5-6 GURT. The three areas/operators/arenas of the GURT are:

1.) The **"Exterior"** = "Outside", symbolized by "**ext**" using Trichotomous Tripositive formatting the exterior is expressed in the following manner—the GURT "Side *a*" exterior is expressed in the following manner, "**ext[a] = 4**". All of the GURT exteriors are illustrated in the graphical model that follows.

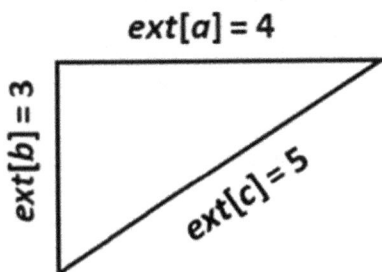

Note, the three Exterior GURT measures are as follows:

ext[a] = Side *a*;
ext[b] = Side *b*;
ext[c] = Side *c*.

Therefore,

ext[a] = Side *a* = 4;
ext[b] = Side *b* = 3;
ext[c] = Side *c* = 5.

2.) The **"Interior"** = "Inside", symbolized by "**int**" using Trichotomous Tripositive formatting the exterior is expressed in the following manner, **int[a]** = "[*ins*[a] for "Side *a*"]". A significant part of the GURT Interior is the "Inscribed Square". The GURT "Inscribed Square" is (in terms of "Side *a*") symbolized as "**ins[a]**" and is expressed mathematically in the following manner, "**ins[a] = 12/7 + 16/7 = 28/7 = 4**":

All of the GURT interiors are illustrated in the three graphical models that follow—

In terms of the Inscribed Square:

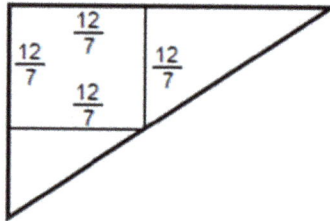

In terms of angles:

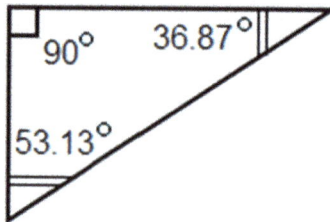

In terms of measurements:

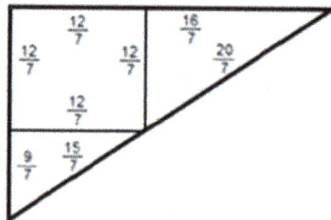

; and

Note, the three exterior GURT measures are as follows:

Interior Angles:

int[∇ac] = "Angle Alpha" ("$\angle \alpha$") = 36.87°;
int[∇ab] = "Angle Rho" ("$\angle \rho$") = 90°;
int[∇bc] = "Angle Beta" ("$\angle \beta$") = 53.13°.

Therefore,

Interior Side Measures:

int[a] = [ins[a] for Inscribed Square "Side a"] = 12/7 + 16/7 = 28/7 = 4;
int[b] = [ins[b] for Inscribed Square "Side b"] = 12/7 + 9/7 = 21/7 = 3;
int[c] = [ins[c] for Inscribed Square "Side c"] = 15/7 + 20/7 = 35/7 = 5.

Lastly, the Interior Inscribed Square measurements are a part of the Interior measurements of the GURT, where "**[s]**" = the "Tripositive Side" measurements of the GURT Inscribed Square:

ins[s] = "The Base Inscribed Square Side Length" = [12/7] = 1.714285714285714285...;
ins[a] = [12/7] + 16/7 = 28/7 = 4 = The complete " ∇ Side a" overall measurement;
ins[b] = [12/7] + 9/7 = 21/7 = 3 = The complete " ∇ Side b" overall measurement;
ins[c] = 15/7 + 20/7 = 35/7 = 4 = The complete " ∇ Side c" overall measurement;

3.) The **"Anterior"** = "In Front of", symbolized by the acronym "*ant*" " using Trichotomous Tripositive formatting the anterior is expresses the geometric trigonometric relationships between each of the GURT sides. The "Tripositive GURT Anterior" for "Side *a* in relation to Side *c*" is the quotient of "Side *a*" divided by "Side *c*". Furthermore, the "Anterior" of the GURT is expressed as the trigonometric cosine function in the following manner, "*ant[a/c]* = ["Side *a*"/"Side *c*"] = 4/5 = 0.8 = The Trigonometric Cosine". All of the GURT Anterior Tripositions are illustrated in the graphical model that follows.

ant[a/c] = ["Side *a*"/"Side *c*"] = 4/5 = 0.8 = The Trigonometric Cosine Function;
ant[b/c] = ["Side *b*"/"Side *c*"] = 3/5 = 0.6 = The Trigonometric Sine Function; and
ant[b/a] = ["Side *b*"/"Side *a*"] = 4/5 = 0.75 = The Trigonometric Tangent Function =
ant[b/a] = ["Side *b*"/"Side *a*"] = 4/5 = 0.75 = The Trichometric Tripositive Inclination.

How Three Geometric Mathematical Fields are Tripositively Trichotomously Related

It is important to note that three "Tripositive Trichotomous Relationships" can be observed in the three main geometric mathematical fields. The trichotomous relationships are written in the following manner:

a.) **"[+]"** The <u>Positive</u> represented by **Trichometry** (syllabized as: "Tri·chom·e·try" and pronounced in the following manner: "/tri'kämətrē/")—in terms of "Tripositive Trichotomy" (which states that for the given positive integers identified as the numbers "*x*" and "*y*", exactly one of the following relations: "[$x < y$]" or "[$x = y$]" or "[$x > y$]", holds), "Trichometry" is the Triological Scientific mathematical field that studies the geometrics of the 3-4-5-6 Golden Upright Right Triangle;

b.) **"[–]"** The <u>Negative</u> represented by **Trigonometry** (syllabized as: "Trig·o·nom·e·try" and pronounced in the following manner: "/trigə'nämətrē/")—in terms of "Tripositive Trichotomy" (which states that for the given positive integers identified as the numbers "*x*" and "*y*", exactly one of the following relations: "[*x* < *y*]" or "[*x* = *y*]" or "[*x* > *y*]", holds), "Trigonometry" is the mathematical field that is not Trichometry that is concerned with measures triangles.; and

c.) **"[ø]"** The <u>Neutral</u> represented by **Triometry** (syllabized as: "Triometry" and pronounced in the following manner: "/tri'ämətrē/")— in terms of "Tripositive Trichotomy" (which states that for the given positive integers identified as the numbers "*x*" and "*y*", exactly one of the following relations: "[*x* < *y*]" or "[*x* = *y*]" or "[*x* > *y*]", holds), "Triometry" is the mathematical field that is neither Trichometry or Trigonometry the mathematical field that is related to Visualus™ © and measures trichotomous digital data archival structures.

Trichometry and Trigonometric Functions

Trigonometric Functions and their properties are expressed in terms Trichometry are written with their full name written in the function. As such Sine is literally written out as: "sine". This indicates that the Trigonometric Function is being used in terms of the mathematics of Trichometry. An example of the use of Trigonometric Functions expressed in terms of the mathematics of Trichometry is the "Cosine Function". The Trigonometric Cosine Function is periodic and expressed in terms of "twice pi", thus, 2π = 6.28318531... The Trigonometry Cosine Function written in terms of Trichometry is written as follows:

The Trigonometric Cosine Function = $(\text{Sine } \theta)^2 + (\text{Cosine } \theta)^2 = 1$.

The abovementioned can be expressed in terms of the "Trichometric Anterior" in the following manner:

ant["Side b"/"Side c"]^2 + ant["Side a"/ "Side c"]^2 = $ant[b/c]^2$ + $ant[a/c]^2$ = $ant[3/5]^2$ + $ant[4/5]^2$ = 1.00.

The aforesaid can also be mathematically expressed as follows:

$ant[b/c]^2$ + $ant[a/c]^2$ = $ant[0.6]^2$ + $ant[0.8]^2$ = $[0.6]^2$ + $[0.8]^2$ = 0.36 + 0.64 = 1.

The Graphical Model that illustrates the aforementioned is as follows:

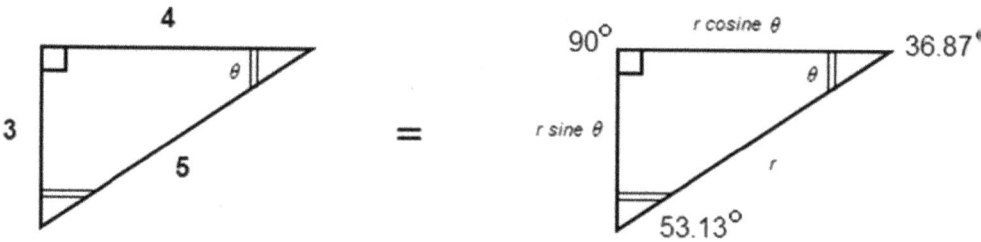

A List of Trigonometric Functions in Terms of Trichometry

The Trichometry Trigonometric Functions Graphical Model

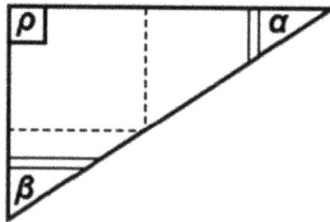

Trigonometry expressed in terms of Trichometry "Intra-Trichotomy" using "Trioengineering Notation" (via " $\triangledown$ " symbol used to denote unique GURT characteristics in terms of Trioengineering):

(1.) The Trichometry of the 3-4-5-6 GURT as a Trigonometric Function of the first angle alpha$_1$—

The Tripositive Trioengineered Sine α = Sine α = [Sine α = Sine θ], therefore, $\triangledown$[Sine θ] (is mathematically defined as: "The Tripositive Trioengineered Trichometry Sine Theta") = $\triangledown$[Sine α] = $\triangledown$["Side b"/"Side c"] = $\triangledown$[3/5] = 0.6;

(2.) The Trichometry of the 3-4-5-6 GURT as a Trigonometric Function of the first angle alpha$_2$—

The Tripositive Cosine α = [Cosine α = Cosine θ], therefore, $\triangledown$[Cosine θ] (is mathematically defined as: "The Tripositive Trioengineered Trichometry Cosine Theta") = $\triangledown$[Cosine α] = $\triangledown$["Side a"/"Side c"] = $\triangledown$[4/5] = 0.8;

(3.) The Trichometry of the 3-4-5-6 GURT as a Trigonometric Function of the second angle beta$_1$—

The Tripositive Sine β = $\triangledown$[Sine β] (is mathematically defined as: "The Tripositive Trioengineered Trichometry Sine Beta") = $\triangledown$["Side a"/"Side c"] = $\triangledown$[4/5] = 0.8;

(4.) The Trichometry of the 3-4-5-6 GURT as a Trigonometric Function of the second angle beta$_2$—

The Tripositive Cosine β = $\triangledown$[Cosine β] (is mathematically defined as: "The Tripositive Trioengineered Trichometry Cosine Beta") = $\triangledown$["Side b"/"Side c"] = $\triangledown$[3/5] = 0.6;

(5.) The Trichometry of the 3-4-5-6 GURT as a Trigonometric Function of the third angle rho_1—

The Tripositive Tangent α = $\sqrt{}$[Tangent α] (is mathematically defined as: "The Tripositive Trioengineered Trichometry Tangent Alpha") = $\sqrt{}$ ["Side b"/"Side a"] = $\sqrt{}$[3/4] = **0.75** = " $\sqrt{}$Inclination" = " $\sqrt{}$Accline", a novel term meaning the "Trioengineered Accline" or " $\sqrt{}$Acclivity" = " $\sqrt{}$ Side Diagonal" (referred to in Trigonometry as the "hypotenuse"); and

(6.) The Trichometry of the 3-4-5-6 GURT as a Trigonometric Function of the third angle rho_2—

The Tripositive Tangent β = $\sqrt{}$[Tangent β] (is mathematically defined as: "The Tripositive Trioengineered Trichometry Tangent Beta") = $\sqrt{}$ ["Side a"/"Side b"] = $\sqrt{}$[4/3] = **1.333...**

The Various Trichometry Ratios, Proportions, and Scalar Measurements in Units, Degrees, and Radians

The following graphical models illustrate Trichometric ratios, proportions, and scalar measurements in units, degrees, and radians in terms of Trigonometry in the graphical models that follow.

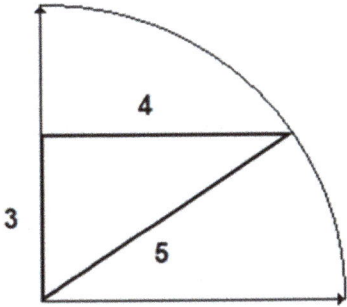

This "Quadracircle" is in the Positive
First Cartesean Coordinate and
measures a grand total of 1.57 rad =
90 degrees.

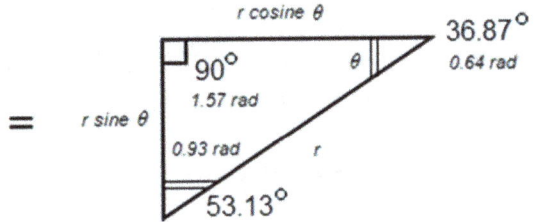

The above graphical models have angle calculations that Mathematically, using radian measurements (that are denoted by using the shortened "rad" abbreviation) the following angular measurements, graphical model, and calculations:

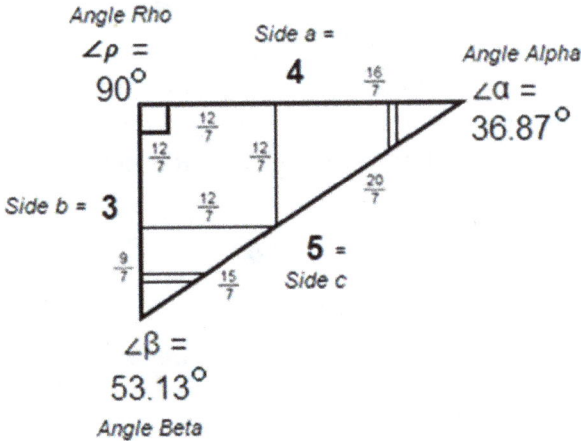

0.64 rad = 36.87° for "angle alpha"
0.93 rad = 53.13° for "angle beta"
+ 1.57 rad = 90°   for "angle rho"
3.14 rad = 180°, where, 180° is the number of degrees for all triangles including the GURT.

Trichometry Utility Illustrated via a "Concrete Example"

The ideal measure of "compound interest" in terms of the 3-4-5-6 GURT (which through the mathematics of Trichometry illustrates the equality of the three most moted mathematical functions and parameters as: "The Trichotomous Tripositive Trioengineered Pi" ("$\pi_{[\triangledown]}$" = "$\pi_{\triangledown}$") = "The Trichotomous Tripositive Trioengineered Phi" ("$\phi_{[\triangledown]}$" = "$\phi_{\triangledown}$") = "The Trichotomous Tripositive Trioengineered Natural Logarithm" ("$e_{[\triangledown]}$" = "$e_{\triangledown}$") that is the "The Trichotomous Tripositive Trioengineered **Measure of Compound Interest" = ["12 Months"/"7 Days of the Week"] + 1 Year = 2.714285714285714285...** The foundations of the aforesaid is illustrated and exhibited in the graphical model that follows and displays the "Inscribed Square" that is a part of the GURT and an "extension of angle rho" that is an active part of defining the "right angle' that formulates GURT and provides it with its unique nature. The angle sides are presented and emphasized in bold to highlight the main areas that enable the GURT to :

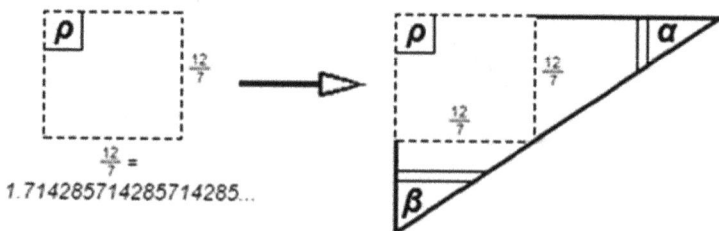

$\frac{12}{7} =$
1.7142857142857714285...

The Intra-Trichotomy of Trichometry

The "Intra-Trichotomy" is defined as the characteristics that are a part of the of the whole 3-4-5-The two 3-4-5-6 Golden Upright Right Triangle as mentioned earlier. The two "Catheti" that are inherent to the GURT aid in defining the outer side lengths that are innate characteristics of ½ of the front face of Visualus Isometric Cuboid. This is easily observed when the GURT and its inverse are placed side by side as illustrated in the series of graphical models that follows.

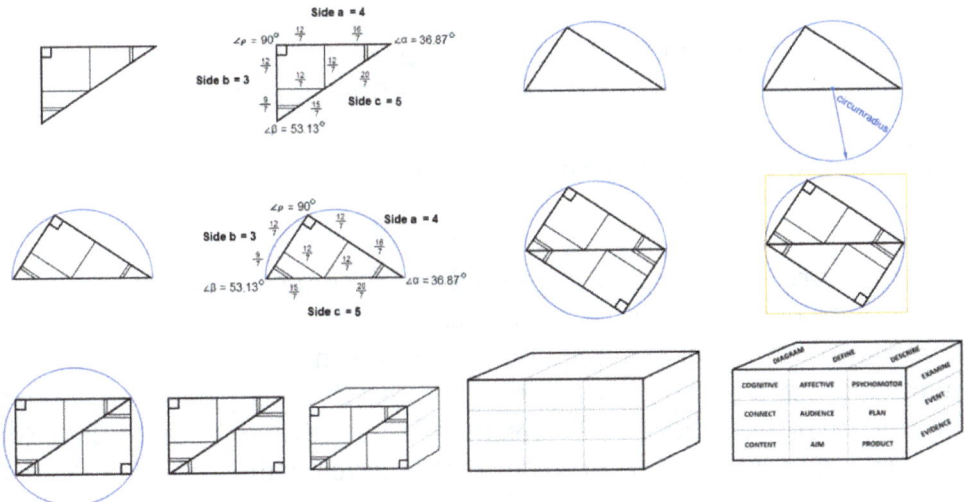

In the displayed graphical model above: The two side lengths of the GURT as the two "Catheti" form the side lengths of the traditional 3 by 3 Tri–Squared Table used in Triostatistics. Additionally, (as illustrated above) this then formulates the "Front Face" of each cell of the "Standard 3 by 3 Tri–Squared Test Table". The "Standard 3 by 3 Tri–Squared Test Table" can then be extended to form the "Front Face" of the Visualus™ © Isometric Cuboid ©.

Trichometric Key Components of the 3-4-5-6 GURT

Pi is equal to 3.14 rad (rad for "radians" in radial measurement). Therefore, all triangles by their innate characteristics are "pi" ("π") because "3.14rad" (or "3.14 rad") = 179.9087° ≅ 180°. As a result, all triangles have an equivalent measurement that is equal to "π" because "π ≅ 180°". It is important to note that all triangle angles add up to 180°. Therefore, all triangles are "pi" or "[π]" is not a mathematical farfetched statement. Note, in terms of radian measurement the following to be true:

Converting [3.14 rad] into degrees ("°"), because:
"[3.14 rad × 180/π = 179.909° ≅ 180°]".

Additionally, pi = C/D, where, C = 'Circumference" and D = "Diameter". Thus,

Furthermore, all triangles are equal to ½ semicircles that are two-coordinate depictions of hemispheres ("½ of a sphere"). Thus, a two-coordinate depiction of a "hemisphere" is a "semicircle" that, (due to its diameter) has a radial measurement of 180° that is equal to the precise number of degrees in any triangle. In turn, as previously stated the 180° of a triangle = 3.14 rad. Therefore, a semicircle (as illustrated in Thales' Theorem) is equal to a triangle in measurement because a two-coordinate hemisphere = a semicircle with a radial measurement of 180° is equal to the measurement in degrees of all triangle angles = 180°. Thus,

"π = ∇ ", and
"π = *All Triangles*" = 180° = Semicircle = 3-4-5-6 GURT;
So, π = Δ (representing "change" in coordinates) *and*
π = ∇ (representing magnitude and direction);
"pi" is equal to Delta and Nabla (or "Del").

It is also important to note that the Inverse of Trigonometry = The Mathematics of Trichometry in terms of the 3-4-5-6 Golden Upright Triangle. In addition, the 3-4-5-6 GURT is the inverse of the traditional right triangle measurements of the mathematics of Trigonometry when applied to a trigonometric right triangle with the overall side length measurements that are 3-4-5 (with an overall Area = 6).

In terms of the mathematics of Trigonometry, the mathematics of Trichometry provides the following:

Length of "Side b" (or "$\bar{b}$") = 3 = the Trigonometric Sine;
Length of "Side a" (or "$\bar{a}$") = 4 = the Trigonometric Cosine; and
Length of "Side c" (or "$\bar{c}$") = 5 = the Trigonometric Hypotenuse.

Chapter Six follows and describes Trichometry Measurements.

I, even I, am [HE] that blotteth out thy transgressions.

Isaiah 43: 25

Trichometry in Action via Foundations: Explaining the Mathematics of Trichotomy

The Mathematical Law of Trichotomy for the Triostatistics Tri–Squared Analysis Statistical Mathematical Model of the Tri–Squared Test = [Tri ^{2}] (extracted from the "Triology" research article—Osler, 2017)

Tri–Square or Tri–Squared comprehensively stands for "The Total Transformative Trichotomous–Squared Test" (or "Trichotomy–Squared"). The Total Transformative Trichotomous–Squared Test provides a methodology for the transformation of the outcomes from qualitative research into measurable quantitative values that are used to test the validity of hypotheses. It is based on the mathematical "Law of Trichotomy". In terms of mathematics, Apostol in his book on calculus defined "The Law of Trichotomy" as: Every real number is negative, 0, or positive. The law is sometimes stated as "For arbitrary real numbers a and b, exactly one of the relations: 1.) $a < b$; 2.) $a = b$; and 3.) $a > b$ holds" (Apostol, 1967). It is important to note that in mathematics, the law (or axiom) of trichotomy is most commonly the statement that for any (real) numbers x and y, exactly one of the following relations holds. Until the end of the 19th century the law of trichotomy was tacitly assumed true without having been thoroughly examined (Singh, 1997).

A proof was sought by Logicians and the law was indeed proved to be true. If applied to cardinal numbers, the law of trichotomy is equivalent to the axiom of choice. More generally, a binary relation R on X is trichotomous if for all x and y in X exactly one of xRy, yRx or $x = y$ holds. If such a relation is also transitive it is a strict total order; this is a special case of a strict weak order. For example, in the case of three elements the relation R given by aRb, aRc, bRc is a strict total order, while the relation R given by the cyclic: 1.) aRb; 2.) bRc; and 3.) cRa is a non–transitive trichotomous relation.

In the definition of an ordered integral domain or ordered field, the law of trichotomy is usually taken as more foundational than the law of total order, with $y = 0$, where 0 is the zero of the integral domain or field. In set theory, trichotomy is most commonly defined as a property that a binary relation < has when all its members $<x, y>$ satisfy exactly one of the relations listed above. Strict inequality is an example of a trichotomous relation in this sense. Trichotomous relations in this sense are irreflexive and antisymmetric (Sensagent, 2012).

The History of Mathematical Law of Trichotomy with foundation of the Tri–Squared Research Design and Explanation of the Tri–Squared Distribution (extracted from the "Triology" research article—Osler, 2017)

The foundational idea of a "Trichotomy" has a detailed long history that is based in discussions surrounding higher cognition, general thought, and descriptions of intellect. Philosopher Immanuel Kant adapted the Thomistic acts of intellect in his trichotomy of higher cognition — (a) understanding, (b) judgment, (c) reason — which he correlated with his adaptation in the soul's capacities — (a) cognitive faculties, (b) feeling of pleasure or displeasure, and (c) faculty of desire (Kant, 2007). The Total Transformative Trichotomous–Squared Test provides a methodology for the transformation of the outcomes from qualitative research into measurable quantitative values that are used to test the validity of hypotheses.

The advantage of this research procedure is that it is a comprehensive holistic testing methodology that is designed to be static way of holistically measuring categorical variables directly applicable to educational and social behavioral environments where the established methods of pure experimental designs are easily violated. The unchanging base of the Tri–Squared Test is the 3 × 3 Table based on Trichotomous Categorical Variables and Trichotomous Outcome Variables. The emphasis the three distinctive variables provide a thorough rigorous robustness to the test that yields enough outcomes to determine if differences truly exist in the environment in which the research takes place (Osler, 2013a).

Tri–Squared is grounded in the combination of the application of the research two mathematical pioneers and the author's research in the basic two-dimensional foundational approaches that ground further explorations into a "tri-coordinate" (i.e., "three-dimensional") Instructional Design. The aforementioned research includes the original dissertation of optical pioneer Ernst Abbe who derived the distribution that would later become known as the chi square distribution and the original research of mathematician Auguste Bravais who pioneered the initial mathematical formula for correlation in his research on observational errors. The Tri–Squared research procedure uses an innovative series of mathematical formulae that do the following as a comprehensive whole: (1) Convert qualitative data into quantitative data; (2) Analyze inputted trichotomous qualitative outcomes; (3) Transform inputted trichotomous qualitative outcomes into outputted quantitative outcomes; and (4) Create a standalone distribution for the analysis possible outcomes and to establish an effective––research effect size and sample size with an associated alpha level to test the validity of an established research hypothesis. The process of designing instruments for the purposes of assessment and evaluation is called "Psychometrics". Psychometrics is broadly defined as the science of psychological assessment (Rust & Golombok, 1989).

The Tri–Squared Test pioneered by the author, factors into the research design a unique event–based "Inventive Investigative Instrument". This is the core of the Trichotomous–Squared Test. The entire procedure is grounded in the qualitative outcomes that are inputted as Trichotomous Categorical Variables based on the Inventive Investigative Instrument (Osler, 2013c). Osler (2012a) initially defined the Tri–Squared mathematical formula in the *Journal on Mathematics* article entitled, "Trichotomy–Squared – A novel mixed methods test and research procedure designed to analyze, transform, and compare qualitative and quantitative data for education scientists who are administrators, practitioners, teachers, and technologists" as follows:

$$Tri^2 = T_{Sum}\left[\left(Tri_x - Tri_y\right)^2 : Tri_y\right]$$

The Tri–Squared distribution is a static mathematical extraction out of the Chi Square distribution. This test is not the only test based on the Chi Square distribution (as it is a mathematical distribution that is frequently used directly or indirectly in many tests of significance). Similar to the Chi Square distribution the Tri–Squared distribution has the following characteristics: (1) It has only a single parameter (the distribution Degrees of Freedom written as "*d.f.*"); (2) The entire distribution is positively skewed; and (3) The Degrees of Freedom are mathematically written, "$[C - 1][R - 1]$" which is equal to the distribution mean. Unlike, the Chi Square distribution the Tri–Squared distribution has the following characteristics:

(1.) The distribution Degrees of Freedom never changes; therefore, it never approaches the Normal Gaussian Distribution (the bell curve);

(2.) As a static test the Tri–Squared Degrees of Freedom is always $[C - 1][R - 1] = [3 - 1][3 - 1] = [2][2] = 4 =$ the distribution mean;

(3.) The distribution mode is always $[d.f. - 2] = [4 - 2] = 2$;

(4.) The distribution median is always approximates [*d.f.* – 0.7] = [4 – 0.7] = 3.3;

(5.) Due to the static or unchanging nature of the distribution, the distribution skew is always positive with the *d.f.* always equaling 4; and

(6.) The distribution formulae uses brackets "[ ]" in its formulaic notations to emphasize "a concentration on" for purposes of clarity.

The Tri–Squared distribution is the foundation for the Tri–Squared Test which comprehensively incorporates the following Tri–Squared formulae: The Calculated Column Standard Deviation, The Calculated Row Standard Deviation, and The Sample Effect Size. The Tri–Squared Test is designed to create a comprehensive holistic research methodology from calculations conducted on the Standard 3 × 3 Tri–Squared Table which produces the following:

(1.) A positive result;

(2.) No information on the variable relationship direction; and

(3.) Associated Effect Size, Sample Size, and Alpha Levels (Osler, 2012b). It is important to note that the research instrument used in Tri–Squared is an invariant (unchanging) fixed static Test.

Describing the Iteration and Recursive Process of Tri–Squared Repeated Measurement (extracted from the "Triology" research article—Osler, 2017)

There are two forms of Repeated Measures in Trichotomously–Squared Inventive Investigative Instruments. They are: 1.) Iterative repetitive Trifold Trichotomous Categorical Variables (a_1, a_2, and a_3); and 2.) Nested Trifold Recursive Trichotomous Outcome Variables (b_1, b_2, and b_3).

"Iteration" is generally defined as the act of process repetition with the aim of reaching a desired target, goal, and/or result. Sequentially each subsequent "iterate" (individual iteration) is a repetition of the process. The outcome of an individual iteration is used as the starting point for the iteration that immediately follows. In the case of Tri–Squared research instruments, the term "Iteration" refers to breakdown of the overall overarching investigation research question into three specific Categorical Variables so that it can be accurately measured. The results of these variables will clearly statistically state whether or not the initial research question has merit.

"Recursion" is broadly defined as the process of repeating items in a self–similar way. For example, of this process consider an illustration that contains multiple or infinite smaller and smaller nested identical images that repetitively occur over and over (as an identical image within an image within an image etc.). The term is applicable to the Tri–Squared researcher designed instrument in that it describes the threefold repetition of the structure of the Trichotomous Categorical Variable sub–questions that are each extracted from the three Categorical Variables (this thereby provides an Inventive Investigative Instrument that has a grand total of nine Trichotomous Outcomes nested within three interrelated but distinctively specific Trichotomous Categorical Variables the tabulated results of which create the Standard 3 × 3 Tri–Squared Table).

The mathematical definition of Trichotomous Repeated Measures in terms of "Iteration" and "Recursion" is represented by the "Trichotomous Invariant Recursive Iterative Formula" written as:

$$Tri_C^2\left[Tri_R^2\right] = 3[3] = 3 \times 3 = \boxed{} = \text{The Standard } 3 \times 3 \text{ Tri–Squared}$$
Table.

Where,

$$Tri_C^2\left[Tri_R^2\right] = \text{Trichotomous–Squared Columns (Categories);}$$

$Tri_C^2[Tri_R^2]$ = Nested Trichotomous–Squared Rows (Outcomes); and

$Tri_C^2[Tri_R^2]$ = Trichotomous Columns (Categories) with Nested Trichotomous Rows (Outcomes) within Trichotomous–Squared Columns.

The aforementioned formula literally means the following: "Trichotomous Outcome Variables in Rows are nested (contained) within Trichotomous Categorical Variable in Columns". Repeated Measures Design is an internal characteristic of the Tri–Squared Test Inventive Investigative Instrument in terms of Trichotomous Outcome variables. The Tri–Squared Test instrument is constructed using the Inventive Investigative Instrument Metric that is Trichotomously Invariant [or "Unchanging"].

Triological Science Tri-Values in terms of Triostatistics Triple Values Expressed in the Tri–Squared Test

The following is extracted from the published research article by the author entitled, "Triology: A Novel, Innovative, and in–Depth Science Concerned with the Mathematical Triadic, Tripartite, and Triplex Components, Content, and Cycles of Life, Learning, Logic, and the Universal Aspects of Nature" published in 2017 in the i-manager's Journal on Mathematics (JMAT) with updated changes that are reflective of the Triological Sciences (all of the associated references that provide the foundation for the aforementioned published research article can be obtained from 2017 JMAT article as the original source).

The "internal invariant triological repeated measures procedures" that are inherent and integral characteristic of Trichotomous Squared Inventive Investigative Instruments is illustrated in the series of Tables that follows in the next section.

Examples of Triostatistics Tri–Squared Test Trichotomous Outcome Variable Repeated Measures in Triostatistics Tri–Squared Tabular Format (extracted from the "Triology" research article—Osler, 2017)

Table 1

The Tri–Squared Test Taxonomy of Trichotomous Outcome Variables [b_1— b_3] Examples of Repeated Measures Terminology

A Taxonomy of Tri–Squared Test Terminology: For the Creation of Inventive Investigative Instruments			
Trichotomous Outcome Variables	Trichotomous Categorical Variables: As Indicators		
	Discipline and Related Content Area: Cognition: *Affective Domain*	Discipline and Related Content Area: Cognition: *Affective Domain*	Discipline and Related Content Area: Chemistry: *Sub-Atomic Particles*
b_1	Yes	Affirmative	Proton
b_2	No	Negative	Electron
b_3	No Opinion	Neutral	Neutron

The abovementioned is expressed as a series of Triostatistics Triangular Equation Models or "[TEM]s" using the Trichometry 3-4-5-6 GURT in the following graphical models:

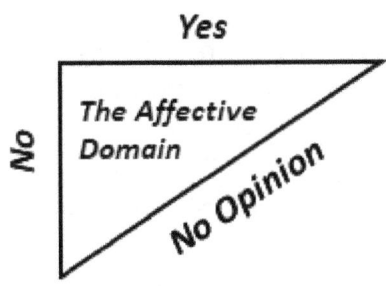

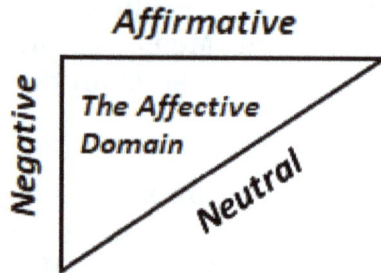

Affirmative — *The Affective Domain* — *Negative* — *Neutral*

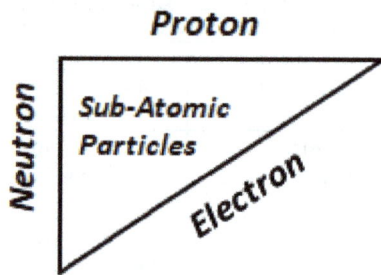

Proton — *Sub-Atomic Particles* — *Neutron* — *Electron*

Table 1 presents the series of Trichotomous Relations that can be used to qualitatively measure as a series of Trichotomous Categorical Variables in the psychological arena of Cognition in the Affective Domain of Learning (in two examples) and the scientific discipline of Chemistry (in one example). The series of Trichotomous Outcome Variables provides (in units of three) the threefold possible "Trichotomous selections" that are the differentiated responses according to the mathematical Law of Trichotomy [as the set of Trichotomous Outcome Variables: b_1; b_2; or b_3]. Table 2 follows and illustrates the next series of possible repeated measures in one single discipline.

Table 2

The Tri–Squared Test Taxonomy of Trichotomous Outcome Variables [$b_1 - b_3$] Examples of Repeated Measures Terminology

A Taxonomy of Tri–Squared Test Terminology: For the Creation of Inventive Investigative Instruments			
	Trichotomous Categorical Variables: As Indicators		
Trichotomous Outcome Variables	Discipline and Related Content Area: Mathematics: *Number Theory*	Discipline and Related Content Area: Mathematics: *Number Theory*	Discipline and Related Content Area: Mathematics: *Volume*
b_1	+	Positive	Full
b_2	–	Negative	Empty
$\underline{b_3}$	∅	Neutral	No Volume

The abovementioned is expressed as a series of Triostatistics Triangular Equation Models or "[TEM]s" using the Trichometry 3-4-5-6 GURT in the following graphical models:

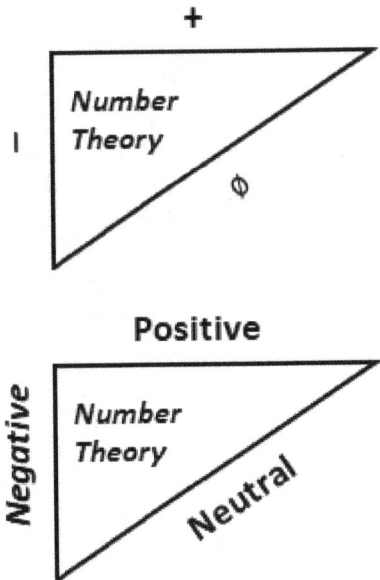

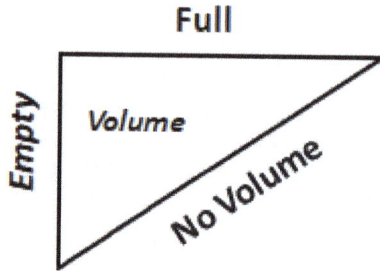

Full

Empty

Volume

No Volume

Table 2 illustrates the series of Trichotomous Relations that can be used to qualitatively measure as a series of Trichotomous Categorical Variables in the academic discipline of Mathematics (in three examples). The series of Trichotomous Outcome Variables provides (in units of three) the threefold possible "Trichotomous selections" that are the differentiated responses according to the mathematical Law of Trichotomy [as the set of Trichotomous Outcome Variables: b_1; b_2; or b_3]. Table 3 follows and illustrates the next series of possible repeated measures in a series of three different disciplines.

Table 3
The Tri–Squared Test Taxonomy of Trichotomous Outcome Variables [$b_1 - b_3$] Examples of Repeated Measures Terminology

A Taxonomy of Tri–Squared Test Terminology: For the Creation of Inventive Investigative Instruments			
Trichotomous Outcome Variables	**Trichotomous Categorical Variables: As Indicators**		
	Discipline and Related Content Area:	Discipline and Related Content Area:	Discipline and Related Content Area:
	Applicable to *All Disciplines*	Engineering: *Electricity*	Psychology: *Decision–Making*
b_1	Positive	High	Agree
b_2	Negative	Low	Disagree
b_3	Neutral	Ground	No Opinion

The abovementioned is expressed as a series of Triostatistics Triangular Equation Models or "[TEM]s" using the Trichometry 3-4-5-6 GURT in the following graphical models:

Positive

Negative | All Disciplines | *Neutral*

High

Low | Electricity | *Ground*

Agree

Disagree | Decision-Making | *No Opinion*

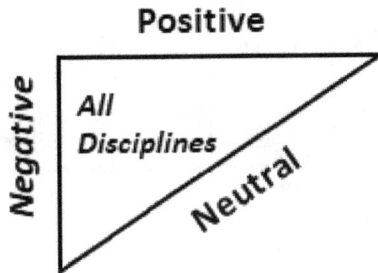

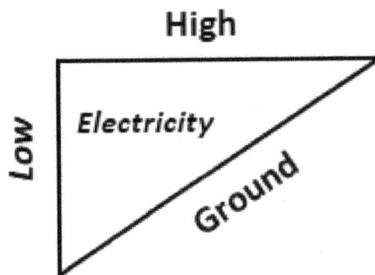

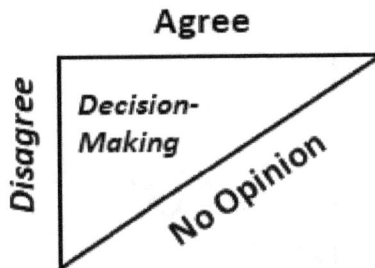

Table 3 displays the series of Trichotomous Relations that can be used to qualitatively measure as a series of Trichotomous Categorical Variables in All Disciplines, the science of Engineering, and the Psychological Arena of Decision–Making (in three different examples). The series of Trichotomous Outcome Variables provides (in units of three) the threefold possible "Trichotomous selections" that are the differentiated responses according to the mathematical Law of Trichotomy [as the set of Trichotomous Outcome Variables: b_1; b_2; or b_3]. Table 4 follows and illustrates the next series of possible repeated measures in a series of three different disciplines.

Table 4

The Tri–Squared Test Taxonomy of Trichotomous Outcome Variables [b_1 — b_3] Examples of Repeated Measures Terminology

A Taxonomy of Tri–Squared Test Terminology: For the Creation of Inventive Investigative Instruments			
Trichotomous Outcome Variables	Trichotomous Categorical Variables: As Indicators		
	Discipline and Related Content Area: *Health Emotional States*	Discipline and Related Content Area: *Mathematics: Cartesian Coordinates*	Discipline and Related Content Area: *Biology: Biometric Identification*
b_1	Happy	Length (x)	Organic
b_2	Sad	Height (y)	Non–Organic
b_3	Calm	Width (z)	No Opinion

The abovementioned is expressed as a series of Triostatistics Triangular Equation Models or "[TEM]s" using the Trichometry 3-4-5-6 GURT in the following graphical models:

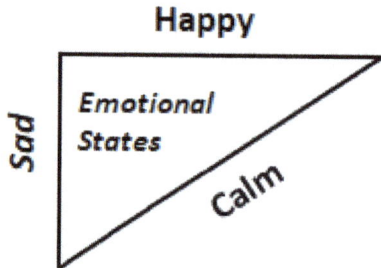

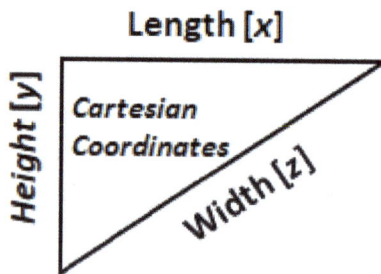

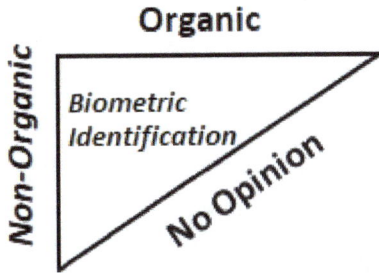

Organic

Non-Organic

Biometric
Identification

No Opinion

Table 4 shows the series of Trichotomous Relations that can be used to qualitatively measure as a series of Trichotomous Categorical Variables in academic discipline of Health, the academic discipline of Mathematics, and the scientific discipline of Biology (in three different examples). The series of Trichotomous Outcome Variables provides (in units of three) the threefold possible "Trichotomous selections" that are the differentiated responses according to the mathematical Law of Trichotomy [as the set of Trichotomous Outcome Variables: b_1; b_2; or b_3]. Table 5 follows and illustrates the next series of possible repeated measures in a series of three different disciplines.

Table 5
The Tri–Squared Test Taxonomy of Trichotomous Outcome Variables [b_1— b_3] Examples of Repeated Measures Terminology

A Taxonomy of Tri–Squared Test Terminology: For the Creation of Inventive Investigative Instruments			
	Trichotomous Categorical Variables: As Indicators		
Trichotomous Outcome Variables	Discipline and Related Content Area:	Discipline and Related Content Area:	Discipline and Related Content Area:
	Learning *Psychomotor Domain*	Quantum Physics: *Particle Motion*	Metaphysics: *Religion (The Soul)*
b_1	Active	Active	Mind
b_2	Still	Inactive	Will
b_3	Asleep	Stasis	Emotions

The aforementioned is expressed as a series of Triostatistics Triangular Equation Models or "[TEM]s" using the Trichometry 3-4-5-6 GURT in the following graphical models:

Active

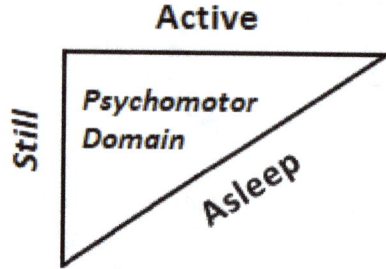

Still | Psychomotor Domain | Asleep

Active

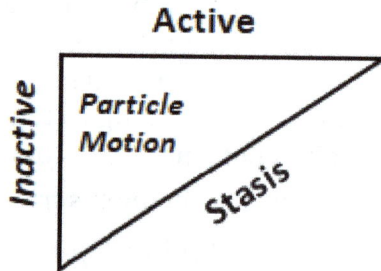

Inactive | Particle Motion | Stasis

Mind

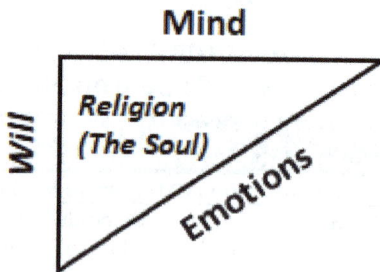

Will | Religion (The Soul) | Emotions

Table 5 expresses the series of Trichotomous Relations that can be used to qualitatively measure as a series of Trichotomous Categorical Variables in the psychological arena of Cognition in the Psychomotor Domain of Learning, the scientific discipline of Quantum Physics, and the spiritual discipline of Metaphysics (in three different examples).

The series of Trichotomous Outcome Variables provides (in units of three) the threefold possible "Trichotomous selections" that are the differentiated responses according to the mathematical Law of Trichotomy [as the set of Trichotomous Outcome Variables: b_1; b_2; or b_3]. Table 6 follows and illustrates the next series of possible repeated measures in a series of three different disciplines.

Table 6
The Tri–Squared Test Taxonomy of Trichotomous Outcome Variables [b_1— b_3] Examples of Repeated Measures Terminology

A Taxonomy of Tri–Squared Test Terminology: For the Creation of Inventive Investigative Instruments			
	Trichotomous Categorical Variables: As Indicators		
Trichotomous Outcome Variables	Discipline and Related Content Area:	Discipline and Related Content Area:	Discipline and Related Content Area:
	Metaphysics: *Religion*	Engineering: *Electricity*	Physics: *Material States*
b_1	Mind	On	Matter
b_2	Body	Off	Energy
b_3	Spirit	No Charge	Vacuum

The abovementioned is expressed as a series of Triostatistics Triangular Equation Models or "[TEM]s" using the Trichometry 3-4-5-6 GURT in the following graphical models:

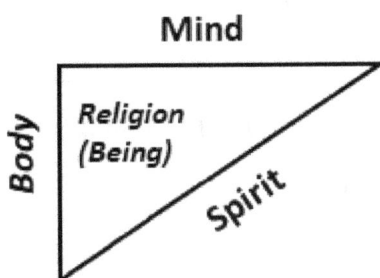

Mind

Body

Religion (Being)

Spirit

On

Off

Electricity
(Engineering)

No Charge

Matter

Energy

Material
States

Vacuum

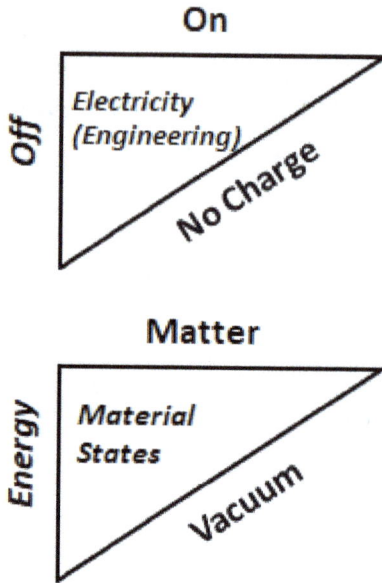

Table 6 states the series of Trichotomous Relations that can be used to qualitatively measure as a series of Trichotomous Categorical Variables in the spiritual discipline of Metaphysics, the scientific discipline of Engineering, and the scientific discipline of Physics (in three different examples). The series of Trichotomous Outcome Variables provides (in units of three) the threefold possible "Trichotomous selections" that are the differentiated responses according to the mathematical Law of Trichotomy [as the set of Trichotomous Outcome Variables: b_1; b_2; or b_3]. Table 7 follows and illustrates the next series of possible repeated measures in a series of three different disciplines.

Table 7

The Tri–Squared Test Taxonomy of Trichotomous Outcome Variables [b_1— b_3] Examples of Repeated Measures Terminology

A Taxonomy of Tri–Squared Test Terminology: For the Creation of Inventive Investigative Instruments			
Trichotomous Outcome Variables	Trichotomous Categorical Variables: As Indicators		
	Discipline and Related Content Area:	Discipline and Related Content Area:	Discipline and Related Content Area:
	Chemistry: *States of Matter*	Engineering: *Bio-Electrical Output*	Mathematics and Computer Science: *Binary Logic*
b_1	Solid	Hard	1
b_2	Liquid	Soft	0
b_3	Gas	Flexible	(empty)

The abovementioned is expressed as a series of Triostatistics Triangular Equation Models or "[TEM]s" using the Trichometry 3-4-5-6 GURT in the following graphical models:

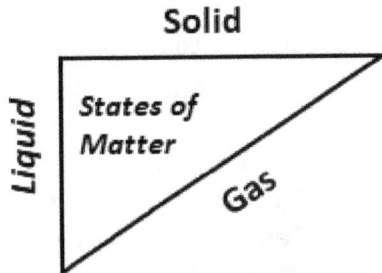

Solid

Liquid

States of Matter

Gas

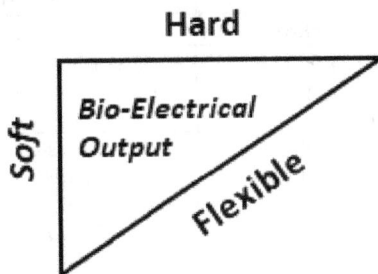

Hard

Soft

Bio-Electrical Output

Flexible

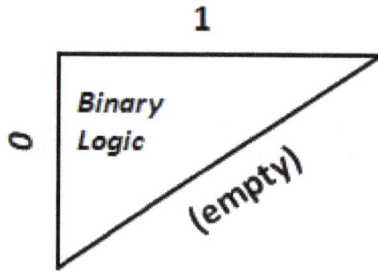

Table 7 confirms the series of Trichotomous Relations that can be used to qualitatively measure as a series of Trichotomous Categorical Variables in the scientific discipline of Chemistry, the scientific discipline of Engineering, and the dual scientific disciplines of Mathematics and Computer Science (in three different examples). The series of Trichotomous Outcome Variables provides (in units of three) the threefold possible "Trichotomous selections" that are the differentiated responses according to the mathematical Law of Trichotomy [as the set of Trichotomous Outcome Variables: b_1; b_2; or b_3]. Table 8 follows and illustrates the next series of possible repeated measures in a series of three different disciplines.

Table 8
The Tri–Squared Test Taxonomy of Trichotomous Outcome Variables [b_1— b_3] Examples of Repeated Measures Terminology

Trichotomous Outcome Variables	A Taxonomy of Tri–Squared Test Terminology: For the Creation of Inventive Investigative Instruments		
	Trichotomous Categorical Variables: As Indicators		
	Discipline and Related Content Area: Geography and Mathematics: *Geospatial Directions*	Discipline and Related Content Area: Metaphysical: *States of Being*	Discipline and Related Content Area: Mathematics: *Set Theory*
b_1	Horizontal	Mental	$\{1, 2, 3...\}$
b_2	Vertical	Physical	$\{..., -3, -2, -1\}$
b_3	Diagonal	Spiritual	$\{0\}$

The abovementioned is expressed as a series of Triostatistics Triangular Equation Models or "[TEM]s" using the Trichometry 3-4-5-6 GURT in the following graphical models:

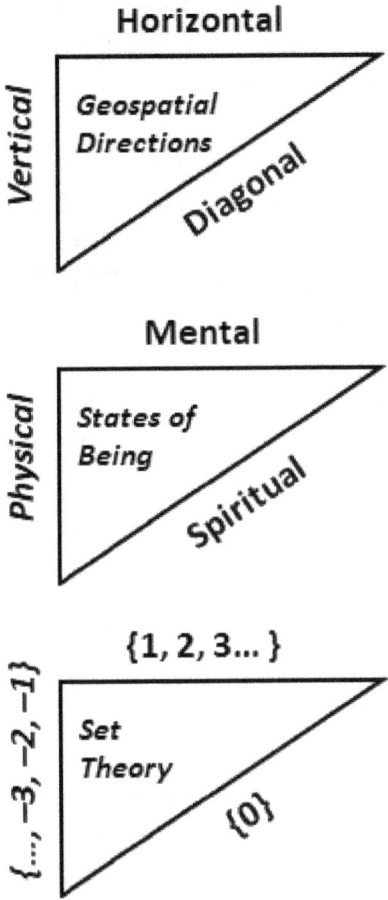

Horizontal

Mental

Table 8 ratifies the series of Trichotomous Relations that can be used to qualitatively measure as a series of Trichotomous Categorical Variables in the dual scientific disciplines of Geography and Mathematics (in terms of Geospatial Relations), the spiritual discipline of Metaphysics, and again in the scientific discipline of Mathematics (in three different examples).

The series of Trichotomous Outcome Variables provides (in units of three) the threefold possible "Trichotomous selections" that are the differentiated responses according to the mathematical Law of Trichotomy [as the set of Trichotomous Outcome Variables: b_1; b_2; or b_3]. Table 9 follows and illustrates the next series of possible repeated measures in a series of three different disciplines.

Table 9
The Tri–Squared Test Taxonomy of Trichotomous Outcome Variables [b_1 — b_3] Examples of Repeated Measures Terminology

Trichotomous Outcome Variables	A Taxonomy of Tri–Squared Test Terminology: For the Creation of Inventive Investigative Instruments		
	Trichotomous Categorical Variables: As Indicators		
	Discipline and Related Content Area:	Discipline and Related Content Area:	Discipline and Related Content Area:
	Mathematics: *Triangulation Coordinates*	Statistics: *Research Accuracy*	Science: *Natural Arrangement*
b_1	x = Abscissa	Objective	Systemic
b_2	y = Ordinate	Biased	Random
b_3	z = Applicate	Uncertain	Unstructured

The abovementioned is expressed as a series of Triostatistics Triangular Equation Models or "[TEM]s" using the Trichometry 3-4-5-6 GURT in the following graphical models:

x = Abscissa

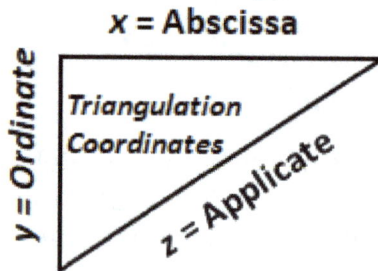

Triangulation Coordinates

y = Ordinate

z = Applicate

Objective

Research
Accuracy

Biased

Uncertain

Systemic

Natural
Arrangement

Random

Unstructured

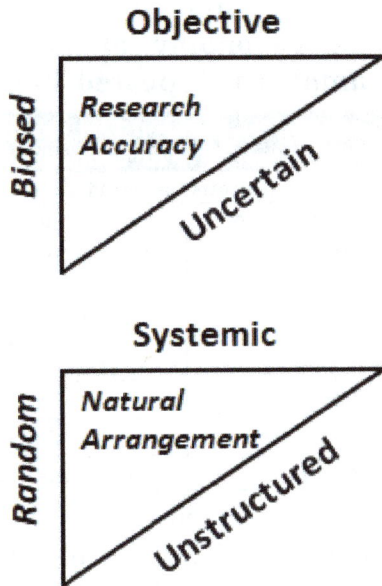

Table 9 illustrates the series of Trichotomous Relations that can be used to qualitatively measure as a series of Trichotomous Categorical Variables in the scientific discipline of Mathematics, the scientific mathematical research–based discipline of Statistics, and in the global general discipline of Science (in three different examples). The series of Trichotomous Outcome Variables provides (in units of three) the threefold possible "Trichotomous selections" that are the differentiated responses according to the mathematical Law of Trichotomy [as the set of Trichotomous Outcome Variables: b_1; b_2; or b_3]. Table 10 follows and illustrates the next series of possible repeated measures in a series of two different disciplines.

Table 10
The Tri–Squared Test Taxonomy of Trichotomous Outcome Variables [b_1— b_3] Examples of Repeated Measures Terminology

A Taxonomy of Tri–Squared Test Terminology: For the Creation of Inventive Investigative Instruments			
Trichotomous Outcome Variables	**Trichotomous Categorical Variables: As Indicators**		
	Discipline and Related Content Area:	Discipline and Related Content Area:	Discipline and Related Content Area:
	Judicial: *Legal Judgment*	Statistics: *Accuracy of Research*	Statistics: *Accuracy of Research*
b_1	Partiality	Valid	Consistent
b_2	Impartiality	Invalid	Random
b_3	Perplexity	Non–Relevant	Unmethodical

The abovementioned is expressed as a series of Triostatistics Triangular Equation Models or "[TEM]s" using the Trichometry 3-4-5-6 GURT in the following graphical models:

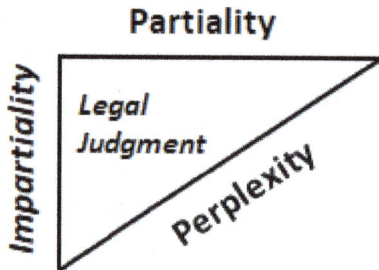

Partiality
Impartiality
Legal Judgment
Perplexity

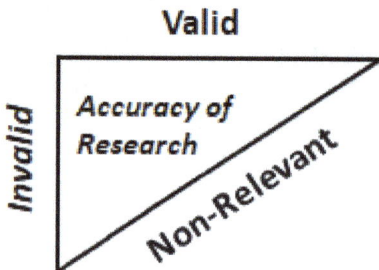

Valid
Invalid
Accuracy of Research
Non-Relevant

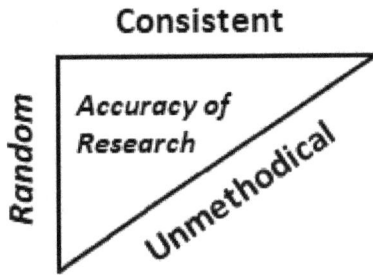

Consistent

Accuracy of Research

Random

Unmethodical

Table 10 displays the series of Trichotomous Relations that can be used to qualitatively measure as a series of Trichotomous Categorical Variables in the legal discipline of the Judiciary, and in the scientific mathematical research–based discipline of Statistics (in three different examples). The series of Trichotomous Outcome Variables provides (in units of three) the threefold possible "Trichotomous selections" that are the differentiated responses according to the mathematical Law of Trichotomy [as the set of Trichotomous Outcome Variables: b_1; b_2; or b_3]. Table 11 follows and illustrates the next series of possible repeated measures in a series of one single discipline.

Table 11

The Tri–Squared Test Taxonomy of Trichotomous Outcome Variables [b_1 — b_3] Examples of Repeated Measures Terminology

A Taxonomy of Tri–Squared Test Terminology: For the Creation of Inventive Investigative Instruments			
	Trichotomous Categorical Variables: As Indicators		
Trichotomous Outcome Variables	Discipline and Related Content Area: Science: *Structure of Experiments*	Discipline and Related Content Area: Science: *Structure of Experiments*	Discipline and Related Content Area: Science: *Structure of Experiments*
b_1	Contingent	Treatment	Predictable
b_2	Unconditional	Outcome	Unpredictable
b_3	Non–Existent	Control	Static

The aforementioned is expressed as a series of Triostatistics Triangular Equation Models or "[TEM]s" using the Trichometry 3-4-5-6 GURT in the following graphical models:

Contingent

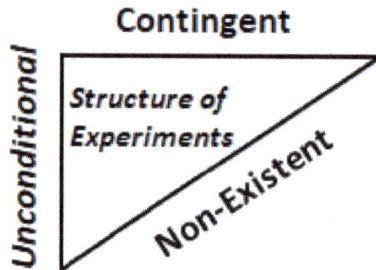

Treatment

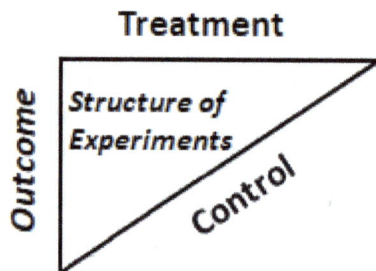

Predictable

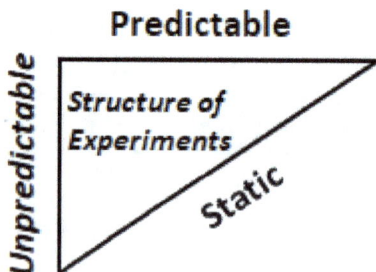

Table 11 explains the series of Trichotomous Relations that can be used to qualitatively measure as a series of Trichotomous Categorical Variables in the global general discipline of Science (in three different examples).

The series of Trichotomous Outcome Variables provides (in units of three) the threefold possible "Trichotomous selections" that are the differentiated responses according to the mathematical Law of Trichotomy [as the set of Trichotomous Outcome Variables: b_1; b_2; or b_3]. Table 12 follows and illustrates the next series of possible repeated measures in a series of three different disciplines.

Table 12

The Tri–Squared Test Taxonomy of Trichotomous Outcome Variables [b_1— b_3] Examples of Repeated Measures Terminology

A Taxonomy of Tri–Squared Test Terminology: For the Creation of Inventive Investigative Instruments			
	Trichotomous Categorical Variables: As Indicators		
Trichotomous Outcome Variables	Discipline and Related Content Area: *Physics and Art: Hue and Light Values*	Discipline and Related Content Area: *Mathematics: Cartesian Coordinates*	Discipline and Related Content Area: *History: Human Affairs*
b_1	Black	Horizontal	Peace
b_2	White	Vertical	War
b_3	Gray	Diagonal	Negotiation

The abovementioned is expressed as a series of Triostatistics Triangular Equation Models or "[TEM]s" using the Trichometry 3-4-5-6 GURT in the following graphical models:

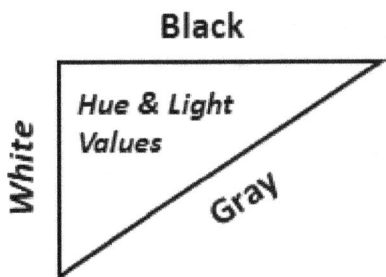

Black

Hue & Light Values

White

Gray

Horizontal

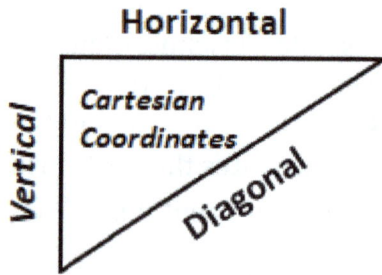

Vertical

Cartesian Coordinates

Diagonal

Peace

War

Cartesian Coordinates

Negotiation

Table 12 clarifies the series of Trichotomous Relations that can be used to qualitatively measure as a series of Trichotomous Categorical Variables in the dual scientific disciplines of Physics and Art, the scientific discipline of Mathematics, and in the academic discipline of History (in three different examples). The series of Trichotomous Outcome Variables provides (in units of three) the threefold possible "Trichotomous selections" that are the differentiated responses according to the mathematical Law of Trichotomy [as the set of Trichotomous Outcome Variables: b_1; b_2; or b_3]. Table 13 follows and illustrates the next series of possible repeated measures in a series of three different disciplines.

Table 13
The Tri–Squared Test Taxonomy of Trichotomous Outcome Variables [b_1— b_3] Examples of Repeated Measures Terminology

A Taxonomy of Tri–Squared Test Terminology: For the Creation of Inventive Investigative Instruments			
	Trichotomous Categorical Variables: As Indicators		
Trichotomous Outcome Variables	Discipline and Related Content Area: Perspective: *Line of Vision*	Discipline and Related Content Area: Art: *Principles of Design*	Discipline and Related Content Area: Mechanics: *Component Connections*
b_1	High	Balanced	Tight
b_2	Low	Unbalanced	Loose
b_3	Horizon	Off Scale	Disconnected

The abovementioned is expressed as a series of Triostatistics Triangular Equation Models or "[TEM]s" using the Trichometry 3-4-5-6 GURT in the following graphical models:

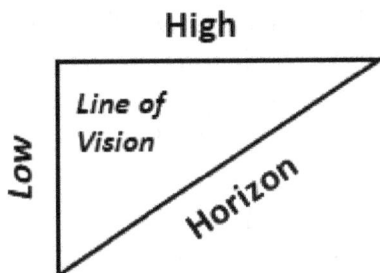

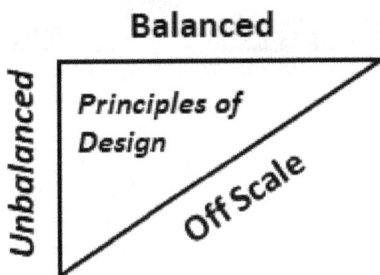

TRICHOMETRY © *The Study of the Geometrics of the 3-4-5-6 Golden Upright Right Triangle in Cartesian Coordinates.* Osler Studios Incorporated ™ © Copyright 2022 All Rights Reserved.

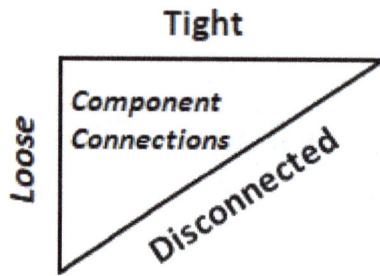

Tight

Loose

Component
Connections

Disconnected

Table 13 elucidates the series of Trichotomous Relations that can be used to qualitatively measure as a series of Trichotomous Categorical Variables in the artistic discipline of Perspective, the academic discipline of Art, and in the specific scientific discipline of general Mechanics (in three different examples). The series of Trichotomous Outcome Variables provides (in units of three) the threefold possible "Trichotomous selections" that are the differentiated responses according to the mathematical Law of Trichotomy [as the set of Trichotomous Outcome Variables: b_1; b_2; or b_3]. Table 14 follows and illustrates the next series of possible repeated measures in a series of one single discipline.

Table 14
The Tri–Squared Test Taxonomy of Trichotomous Outcome Variables [b_1— b_3] Examples of Repeated Measures Terminology

A Taxonomy of Tri–Squared Test Terminology: For the Creation of Inventive Investigative Instruments			
	Trichotomous Categorical Variables: As Indicators		
Trichotomous Outcome Variables	Discipline and Related Content Area:	Discipline and Related Content Area:	Discipline and Related Content Area:
	Mathematics: *Geometric Shapes*	Mathematics: *Primary Operations*	Mathematics: *Primary Operations*
b_1	Square	Addition	Multiplication
b_2	Circle	Subtraction	Division
b_3	Rhombus	Exponentiation	Absolute Value

The aforementioned is expressed as a series of Triostatistics Triangular Equation Models or "[TEM]s" using the Trichometry 3-4-5-6 GURT in the following graphical models:

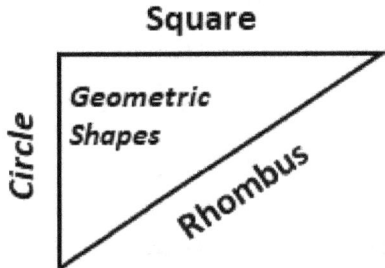

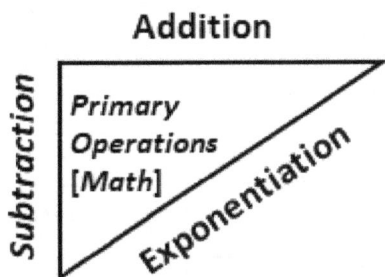

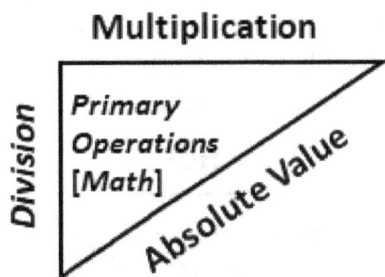

Square

Circle | Geometric Shapes

Rhombus

Addition

Subtraction | Primary Operations [Math]

Exponentiation

Multiplication

Division | Primary Operations [Math]

Absolute Value

Table 14 exemplifies the series of Trichotomous Relations that can be used to qualitatively measure as a series of Trichotomous Categorical Variables in the scientific discipline of Mathematics (in three different examples).

The series of Trichotomous Outcome Variables provides (in units of three) the threefold possible "Trichotomous selections" that are the differentiated responses according to the mathematical Law of Trichotomy [as the set of Trichotomous Outcome Variables: b_1; b_2; or b_3]. Table 15 follows and illustrates the next series of possible repeated measures in a series of two different disciplines.

Table 15
The Tri–Squared Test Taxonomy of Trichotomous Outcome Variables [b_1— b_3] Examples of Repeated Measures Terminology

A Taxonomy of Tri–Squared Test Terminology: For the Creation of Inventive Investigative Instruments			
Trichotomous Outcome Variables	Trichotomous Categorical Variables: As Indicators		
	Discipline and Related Content Area:	Discipline and Related Content Area:	Discipline and Related Content Area:
	Temperature: *Amount of Heat*	Physics: *States of Matter*	Physics: *States of Matter*
b_1	Hot	Hard	Porous
b_2	Cold	Soft	Non–Porous
b_3	Lukewarm	Flexible	Pliable

The abovementioned is expressed as a series of Triostatistics Triangular Equation Models or "[TEM]s" using the Trichometry 3-4-5-6 GURT in the following graphical models:

Hot

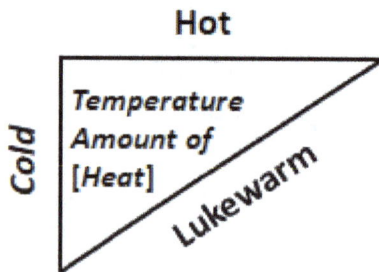

Cold

Temperature Amount of [Heat]

Lukewarm

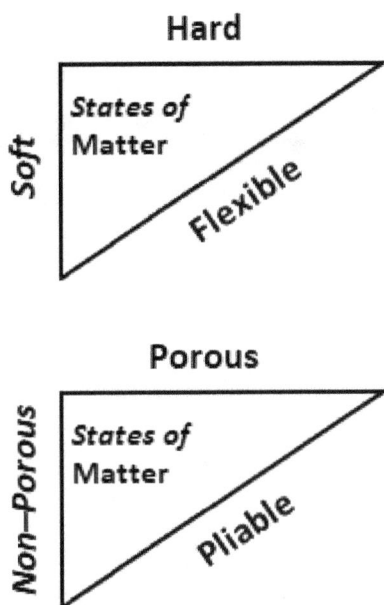

Hard

Soft | States of Matter

Flexible

Porous

Non–Porous | States of Matter

Pliable

Table 15 presents the series of Trichotomous Relations that can be used to qualitatively measure as a series of Trichotomous Categorical Variables in the scientific thermodynamic discipline of Temperature, and in the scientific discipline of Physics (in three different examples). The series of Trichotomous Outcome Variables provides (in units of three) the threefold possible "Trichotomous selections" that are the differentiated responses according to the mathematical Law of Trichotomy [as the set of Trichotomous Outcome Variables: b_1; b_2; or b_3]. Table 16 follows and illustrates the next series of possible repeated measures in a series of two disciplines.

Table 16

The Tri–Squared Test Taxonomy of Trichotomous Outcome Variables [b_1— b_3] Examples of Repeated Measures Terminology

A Taxonomy of Tri–Squared Test Terminology: For the Creation of Inventive Investigative Instruments			
Trichotomous Outcome Variables	Trichotomous Categorical Variables: As Indicators		
	Discipline and Related Content Area: Physics: *Sub-Atomic Movement*	Discipline and Related Content Area: Physics: *States of Matter*	Discipline and Related Content Area: Physics: *States of Matter*
b_1	Active	Dense	Permeable
b_2	Still	Liquid	Non–Penetrable
b_3	Explosive	Gelatinous	Radiant

The abovementioned is expressed as a series of Triostatistics Triangular Equation Models or "[TEM]s" using the Trichometry 3-4-5-6 GURT in the following graphical models:

Active

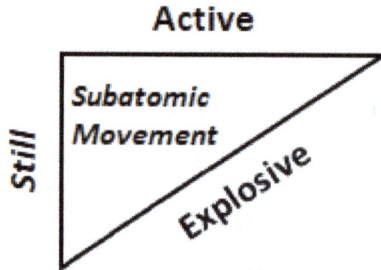

Dense

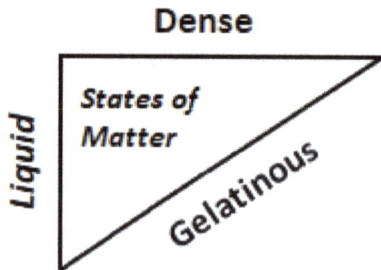

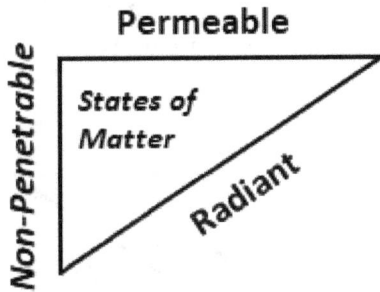

Permeable

Non-Penetrable

States of Matter

Radiant

Table 16 confirms the series of Trichotomous Relations that can be used to qualitatively measure as a series of Trichotomous Categorical Variables in the scientific thermodynamic discipline of Temperature, and in the scientific discipline of Physics (in three different examples). The series of Trichotomous Outcome Variables provides (in units of three) the threefold possible "Trichotomous selections" that are the differentiated responses according to the mathematical Law of Trichotomy [as the set of Trichotomous Outcome Variables: b_1; b_2; or b_3]. Table 17 follows and illustrates the next series of possible repeated measures in a series of two disciplines.

Table 17
The Tri–Squared Test Taxonomy of Trichotomous Outcome Variables [$b_1 - b_3$] Examples of Repeated Measures Terminology

A Taxonomy of Tri–Squared Test Terminology: For the Creation of Inventive Investigative Instruments			
	Trichotomous Categorical Variables: As Indicators		
Trichotomous Outcome Variables	Discipline and Related Content Area:	Discipline and Related Content Area:	Discipline and Related Content Area:
	Psychology: *Learning Domains*	Science: *Structure of Experiments*	Psychology: *Emotional State*
b_1	Cognitive	Independent Variable	Happy
b_2	Affect	Dependent Variable	Sad
b_3	Psychomotor	Control Variable	Calm

The aforementioned is expressed as a series of Triostatistics Triangular Equation Models or "[TEM]s" using the Trichometry 3-4-5-6 GURT in the following graphical models:

Cognitive

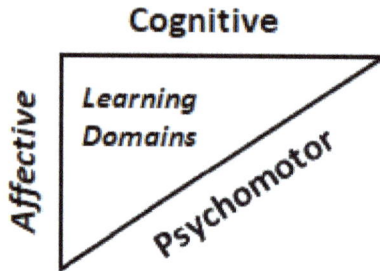

Independent

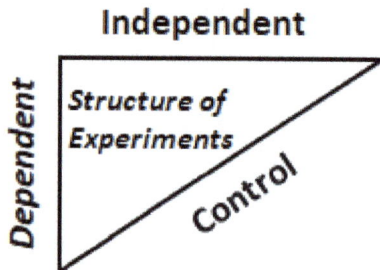

Happy

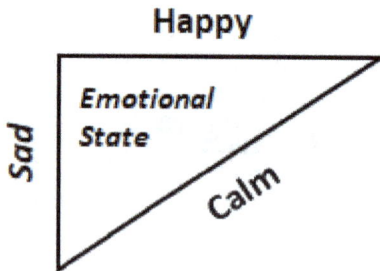

Table 17 shows the series of Trichotomous Relations that can be used to qualitatively measure as a series of Trichotomous Categorical Variables in the academic and scientific discipline of Psychology (twice), and in the general global discipline of Science (in three different examples).

The series of Trichotomous Outcome Variables provides (in units of three) the threefold possible "Trichotomous selections" that are the differentiated responses according to the mathematical Law of Trichotomy [as the set of Trichotomous Outcome Variables: b_1; b_2; or b_3]. Table 18 follows and illustrates the next series of possible repeated measures in a series of three disciplines.

Table 18
The Tri–Squared Test Taxonomy of Trichotomous Outcome Variables [b_1— b_3] Examples of Repeated Measures Terminology

	A Taxonomy of Tri–Squared Test Terminology: For the Creation of Inventive Investigative Instruments		
	Trichotomous Categorical Variables: As Indicators		
Trichotomous Outcome Variables	Discipline and Related Content Area:	Discipline and Related Content Area:	Discipline and Related Content Area:
	Attainment: *Compensation as Merit*	Service: *Helpful Relief and Support*	Geography: *Existing Natural Habitat*
b_1	Rewarded	Helpful	Native
b_2	Unrewarded	Needful	Non–Native
b_3	Non–Participant	Independent	Newcomer

The abovementioned is expressed as a series of Triostatistics Triangular Equation Models or "[TEM]s" using the Trichometry 3-4-5-6 GURT in the following graphical models:

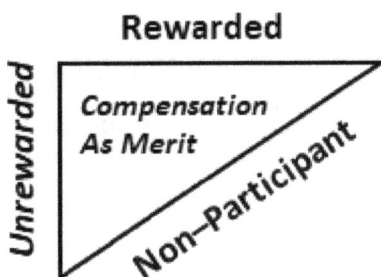

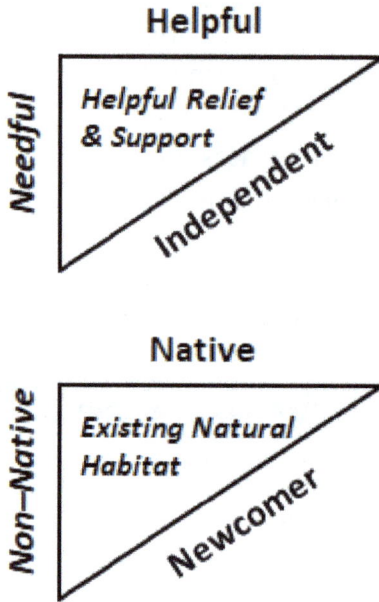

Helpful

Helpful Relief & Support

Independent

Needful

Native

Existing Natural Habitat

Newcomer

Non–Native

Table 18 clarifies the series of Trichotomous Relations that can be used to qualitatively measure as a series of Trichotomous Categorical Variables in the psychometric discipline of Attainment, the community–based discipline of Service, and in the scientific discipline of Geography (in three different examples). The series of Trichotomous Outcome Variables provides (in units of three) the threefold possible "Trichotomous selections" that are the differentiated responses according to the mathematical Law of Trichotomy [as the set of Trichotomous Outcome Variables: b_1; b_2; or b_3]. Table 19 follows and illustrates the next series of possible repeated measures in a series of two different disciplines.

Table 19
The Tri–Squared Test Taxonomy of Trichotomous Outcome Variables [b_1 — b_3] Examples of Repeated Measures Terminology

A Taxonomy of Tri–Squared Test Terminology: For the Creation of Inventive Investigative Instruments			
Trichotomous Outcome Variables	**Trichotomous Categorical Variables: As Indicators**		
	Discipline and Related Content Area:	Discipline and Related Content Area:	Discipline and Related Content Area:
	Metaphysics: *Nature of Existence*	Archaeology: *State of Findings*	Archaeology: *State of Findings*
b_1	Unlimited	Novel	New
b_2	Limited	Ancient	Old
b_3	Undefined	Unknown	Unique

The abovementioned is expressed as a series of Triostatistics Triangular Equation Models or "[TEM]s" using the Trichometry 3-4-5-6 GURT in the following graphical models:

Unlimited

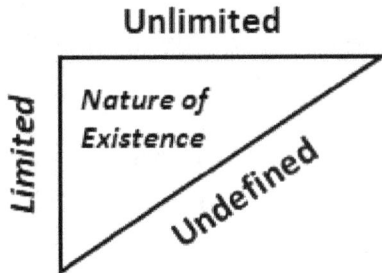

Novel

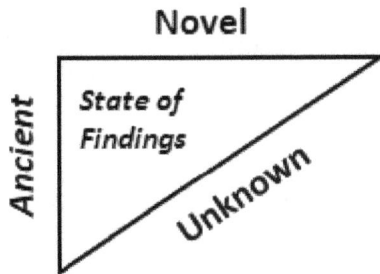

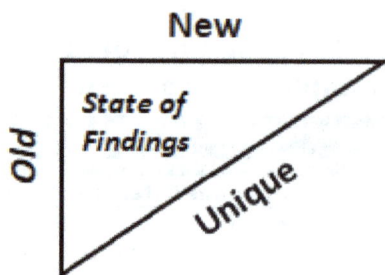

New

State of Findings

Old

Unique

Table 19 shows the series of Trichotomous Relations that can be used to qualitatively measure as a series of Trichotomous Categorical Variables in the spiritual discipline of Metaphysics, and in the academic and scientific discipline of Archaeology (in three different examples). The series of Trichotomous Outcome Variables provides (in units of three) the threefold possible "Trichotomous selections" that are the differentiated responses according to the mathematical Law of Trichotomy [as the set of Trichotomous Outcome Variables: b_1; b_2; or b_3]. Table 20 follows and illustrates the next series of possible repeated measures in a series of three disciplines.

Table 20

The Tri–Squared Test Taxonomy of Trichotomous Outcome Variables [b_1— b_3] Examples of Repeated Measures Terminology

A Taxonomy of Tri–Squared Test Terminology: For the Creation of Inventive Investigative Instruments			
	Trichotomous Categorical Variables: As Indicators		
Trichotomous Outcome Variables	Discipline and Related Content Area:	Discipline and Related Content Area:	Discipline and Related Content Area:
	Quality Control: *Level of Qualification*	Credentialing: *Level of Qualification*	Modeling: *Application Descriptors*
b_1	Compliant	Accredited	Universal
b_2	Non–Compliant	Non–Accredited	Non–Universal
b_3	Non–Credentialed	Unrecognized	Limited

The abovementioned is expressed as a series of Triostatistics Triangular Equation Models or "[TEM]s" using the Trichometry 3-4-5-6 GURT in the following graphical models:

Compliant

Accredited

Universal

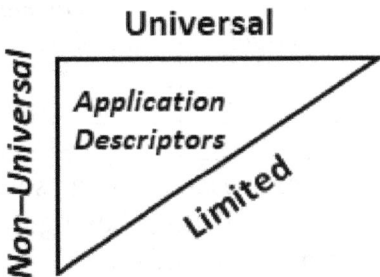

Table 20 elucidates the series of Trichotomous Relations that can be used to qualitatively measure as a series of Trichotomous Categorical Variables in the scientific, ergonomic, and assurance discipline of Quality Control, the academic and expertise discipline of Credentialing, and in the general scientific discipline of applied Modeling (in three different examples). The series of Trichotomous Outcome Variables provides (in units of three) the threefold possible "Trichotomous selections" that are the differentiated responses according to the mathematical Law of Trichotomy [as the set of Trichotomous Outcome Variables: b_1; b_2; or b_3]. Table 21 follows and illustrates the next series of possible repeated measures in a series of three different disciplines.

Table 21
The Tri–Squared Test Taxonomy of Trichotomous Outcome Variables [b_1— b_3] Examples of Repeated Measures Terminology

A Taxonomy of Tri–Squared Test Terminology: For the Creation of Inventive Investigative Instruments			
	Trichotomous Categorical Variables: As Indicators		
Trichotomous Outcome Variables	Discipline and Related Content Area:	Discipline and Related Content Area:	Discipline and Related Content Area:
	Credentialing: *Level of Qualification*	Psychology: *Emotional State*	Modeling: *Nature of Lemma*
b_1	Qualified	Patient	Universal
b_2	Non–Qualified	Impatient	Non–Universal
b_3	Uninterested	Non–Present	Non–Seeking

The abovementioned is expressed as a series of Triostatistics Triangular Equation Models or "[TEM]s" using the Trichometry 3-4-5-6 GURT in the following graphical models:

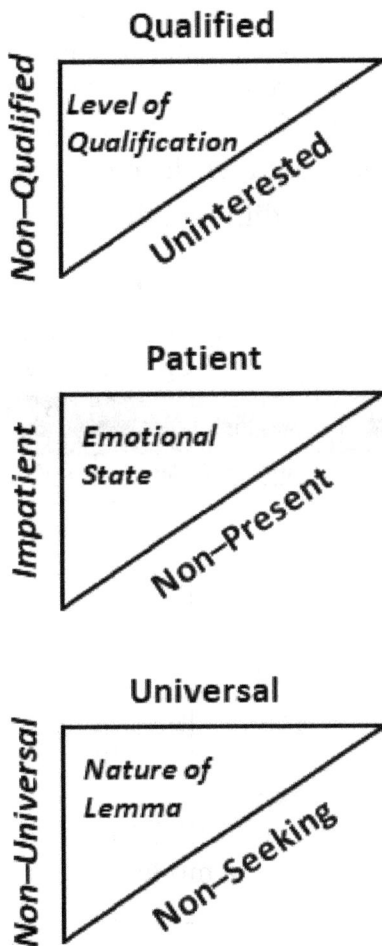

Qualified

Non–Qualified

Level of
Qualification

Uninterested

Patient

Impatient

Emotional
State

Non–Present

Universal

Non–Universal

Nature of
Lemma

Non–Seeking

Table 21 reveals the series of Trichotomous Relations that can be used to qualitatively measure as a series of Trichotomous Categorical Variables in the academic and expertise discipline of Credentialing, the academic and scientific discipline of Psychology, and in the general scientific discipline of applied Modeling (in three different examples).

The series of Trichotomous Outcome Variables provides (in units of three) the threefold possible "Trichotomous selections" that are the differentiated responses according to the mathematical Law of Trichotomy [as the set of Trichotomous Outcome Variables: b_1; b_2; or b_3]. Table 22 follows and illustrates the next series of possible repeated measures in a series of three different disciplines.

Table 22
The Tri–Squared Test Taxonomy of Trichotomous Outcome Variables [b_1— b_3] Examples of Repeated Measures Terminology

	A Taxonomy of Tri–Squared Test Terminology: For the Creation of Inventive Investigative Instruments		
Trichotomous Outcome Variables	Trichotomous Categorical Variables: As Indicators		
	Discipline and Related Content Area:	Discipline and Related Content Area:	Discipline and Related Content Area:
	Physics: *State of Energy*	Psychology: *Concentration Level*	Metaphysics: *State of Being*
b_1	Harnessed	Attentive	Free
b_2	Unharnessed	Distracted	Bound
b_3	Wild	Detached	Delivered

The abovementioned is expressed as a series of Triostatistics Triangular Equation Models or "[TEM]s" using the Trichometry 3-4-5-6 GURT in the following graphical models:

Harnessed

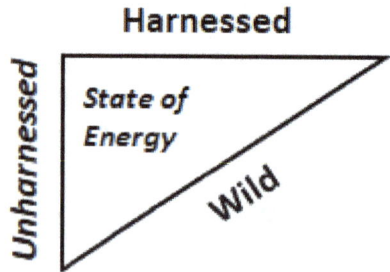

190

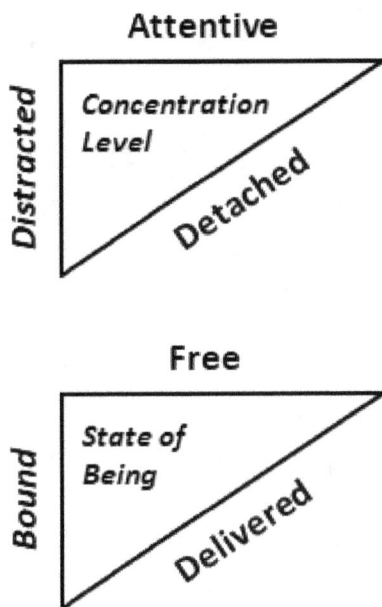

Attentive

Concentration
Level

Distracted

Detached

Free

State of
Being

Bound

Delivered

Table 22 expounds on the series of Trichotomous Relations that can be used to qualitatively measure as a series of Trichotomous Categorical Variables in the academic and scientific discipline of Physics, the academic and scientific discipline of Psychology, and in the spiritual discipline of Metaphysics (in three different examples). The series of Trichotomous Outcome Variables provides (in units of three) the threefold possible "Trichotomous selections" that are the differentiated responses according to the mathematical Law of Trichotomy [as the set of Trichotomous Outcome Variables: b_1; b_2; or b_3]. Table 23 follows and illustrates the next series of possible repeated measures in a series of three different disciplines.

Table 23

The Tri–Squared Test Taxonomy of Trichotomous Outcome Variables [b_1— b_3] Examples of Repeated Measures Terminology

Trichotomous Outcome Variables	A Taxonomy of Tri–Squared Test Terminology: For the Creation of Inventive Investigative Instruments		
	Trichotomous Categorical Variables: As Indicators		
	Discipline and Related Content Area:	Discipline and Related Content Area:	Discipline and Related Content Area:
	Experience: *Level of Expertise*	Physics: *State of Existence*	Nature: *State of Existence*
b_1	Basic	Unity	Occupy
b_2	Intermediate	Chaos	Unoccupied
b_3	Advanced	Nothing	Empty

The abovementioned is expressed as a series of Triostatistics Triangular Equation Models or "[TEM]s" using the Trichometry 3-4-5-6 GURT in the following graphical models:

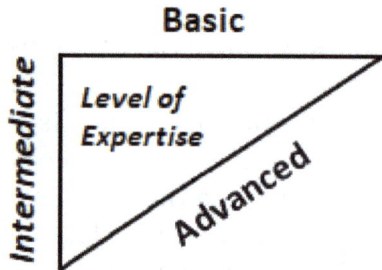

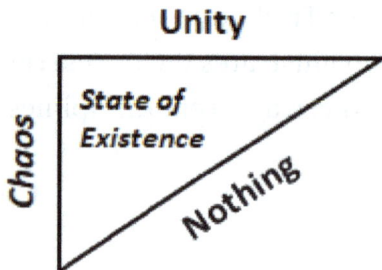

Basic

Level of Expertise

Advanced

Intermediate

Unity

State of Existence

Nothing

Chaos

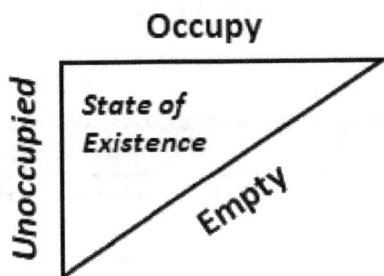

Occupy

State of Existence

Unoccupied

Empty

Table 23 explains the series of Trichotomous Relations that can be used to qualitatively measure as a series of Trichotomous Categorical Variables in the general and global discipline of Experience, the academic and scientific discipline of Physics, and in the global and holistic discipline of Nature (in three different examples). The series of Trichotomous Outcome Variables provides (in units of three) the threefold possible "Trichotomous selections" that are the differentiated responses according to the mathematical Law of Trichotomy [as the set of Trichotomous Outcome Variables: b_1; b_2; or b_3]. Table 24 follows and illustrates the next series of possible repeated measures in a series of three different disciplines.

Table 24
The Tri–Squared Test Taxonomy of Trichotomous Outcome Variables [$b_1 - b_3$] Examples of Repeated Measures Terminology

A Taxonomy of Tri–Squared Test Terminology: For the Creation of Inventive Investigative Instruments			
Trichotomous Outcome Variables	Trichotomous Categorical Variables: As Indicators		
	Discipline and Related Content Area:	Discipline and Related Content Area:	Discipline and Related Content Area:
	Metaphysics: *State of Being*	Science: *Structure of a Substance*	Science: *Arrangement of a Substance*
b_1	Harmony	Systemic	Sequential
b_2	Disharmony	Random	Unorganized
b_3	Quiet	Unknown	Vacant

The abovementioned is expressed as a series of Triostatistics Triangular Equation Models or "[TEM]s" using the Trichometry 3-4-5-6 GURT in the following graphical models:

Harmony

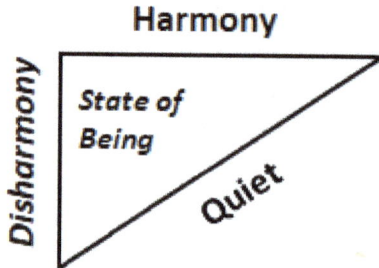

Sequential

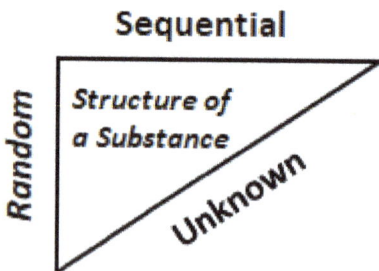

Sequential

Unorganized

Arrangement of a Substance

Vacant

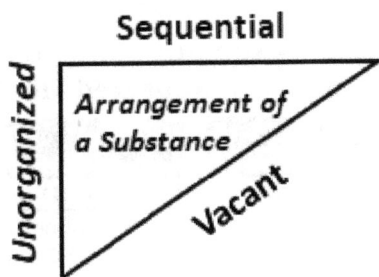

Table 24 confirms the series of Trichotomous Relations that can be used to qualitatively measure as a series of Trichotomous Categorical Variables in the spiritual discipline of Metaphysics and in the general discipline of Science (in three different examples). The series of Trichotomous Outcome Variables provides (in units of three) the threefold possible "Trichotomous selections" that are the differentiated responses according to the mathematical Law of Trichotomy [as the set of Trichotomous Outcome Variables: b_1; b_2; or b_3]. Table 25 follows and illustrates the next series of possible repeated measures in a series of three disciplines.

Table 25

The Tri–Squared Test Taxonomy of Trichotomous Outcome Variables [b_1 — b_3] Examples of Repeated Measures Terminology

A Taxonomy of Tri–Squared Test Terminology: For the Creation of Inventive Investigative Instruments			
Trichotomous Outcome Variables	Trichotomous Categorical Variables: As Indicators		
	Discipline and Related Content Area: Negotiation: *Affairs of State*	Discipline and Related Content Area: Psychometrics: *Decision–Making*	Discipline and Related Content Area: System Structure: *Organizational Dynamics*
b_1	Agreement	Agree	Meeting
b_2	Disagreement	Disagree	No Meeting
b_3	Inconclusive	No Decision	Unscheduled

The abovementioned is expressed as a series of Triostatistics Triangular Equation Models or "[TEM]s" using the Trichometry 3-4-5-6 GURT in the following graphical models:

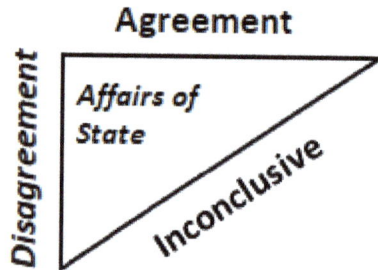

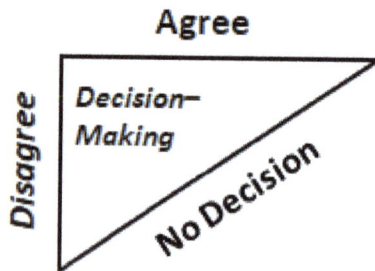

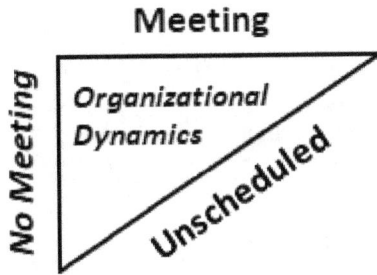

Meeting

No Meeting

Organizational Dynamics

Unscheduled

Table 25 illustrates the series of Trichotomous Relations that can be used to qualitatively measure as a series of Trichotomous Categorical Variables in the business and industry discipline of Negotiation, the academic and scientific discipline of Psychometrics, and in the organizational discipline of the Structure of Systems (in three different examples). The series of Trichotomous Outcome Variables provides (in units of three) the threefold possible "Trichotomous selections" that are the differentiated responses according to the mathematical Law of Trichotomy [as the set of Trichotomous Outcome Variables: b_1; b_2; or b_3]. Table 26 follows and illustrates the next series of possible repeated measures in a series of three different disciplines.

Table 26
The Tri–Squared Test Taxonomy of Trichotomous Outcome Variables [b_1— b_3] Examples of Repeated Measures Terminology

A Taxonomy of Tri–Squared Test Terminology: For the Creation of Inventive Investigative Instruments			
Trichotomous Outcome Variables	Trichotomous Categorical Variables: As Indicators		
	Discipline and Related Content Area:	Discipline and Related Content Area:	Discipline and Related Content Area:
	Awareness: *State of Existence*	System Structure: *Organizational Dynamics*	Science: *Chronological States*
b_1	Here	Group	Timed
b_2	There	No Group	Unrecorded
b_3	Nowhere	Regroup	Paused

The abovementioned is expressed as a series of Triostatistics Triangular Equation Models or "[TEM]s" using the Trichometry 3-4-5-6 GURT in the following graphical models:

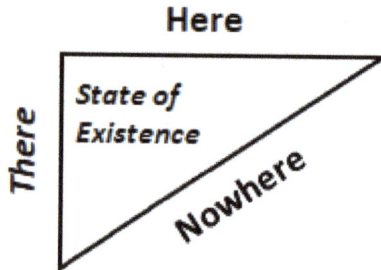

Here

There   *State of Existence*   *Nowhere*

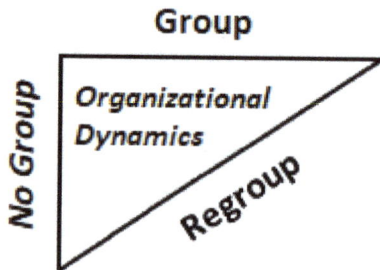

Group

No Group   *Organizational Dynamics*   *Regroup*

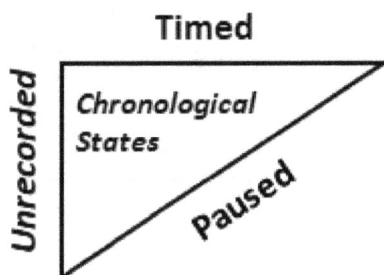

Timed

Chronological States

Unrecorded

Paused

Table 26 verifies the series of Trichotomous Relations that can be used to qualitatively measure as a series of Trichotomous Categorical Variables in the holistic discipline of Awareness, the organizational discipline of the Structure of Systems, and in the general discipline of Science (in three different examples). The series of Trichotomous Outcome Variables provides (in units of three) the threefold possible "Trichotomous selections" that are the differentiated responses according to the mathematical Law of Trichotomy [as the set of Trichotomous Outcome Variables: b_1; b_2; or b_3]. Table 27 follows and illustrates the next series of possible repeated measures in a series of two different disciplines.

Table 27
The Tri–Squared Test Taxonomy of Trichotomous Outcome Variables [b_1— b_3] Examples of Repeated Measures Terminology

A Taxonomy of Tri–Squared Test Terminology: For the Creation of Inventive Investigative Instruments			
	Trichotomous Categorical Variables: As Indicators		
Trichotomous Outcome Variables	Discipline and Related Content Area: *Analysis: Process of Inquiry*	Discipline and Related Content Area: *Analysis: Process of Inquiry*	Discipline and Related Content Area: Science: *Rational States*
b_1	Who	How	Logic
b_2	What	Where	Confusion
b_3	When	Why	Placid

The aforementioned is expressed as a series of Triostatistics Triangular Equation Models or "[TEM]s" using the Trichometry 3-4-5-6 GURT in the following graphical models:

Who

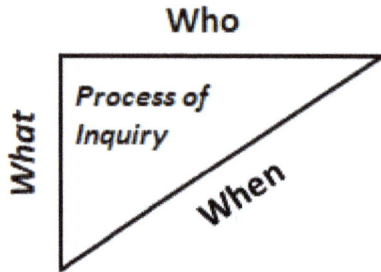

What / *Process of Inquiry* / *When*

How

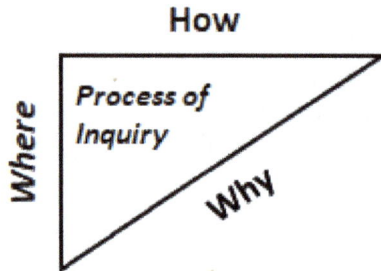

Where / *Process of Inquiry* / *Why*

Logic

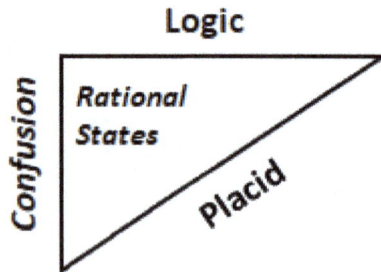

Confusion / *Rational States* / *Placid*

Table 27 provides evidence of the series of Trichotomous Relations that can be used to qualitatively measure as a series of Trichotomous Categorical Variables in the universal scientific discipline of Analysis and in the general discipline of Science (in three different examples).

The series of Trichotomous Outcome Variables provides (in units of three) the threefold possible "Trichotomous selections" that are the differentiated responses according to the mathematical Law of Trichotomy [as the set of Trichotomous Outcome Variables: b_1; b_2; or b_3]. Table 28 follows and illustrates the next series of possible repeated measures in a series of a single individual discipline.

Table 28
The Tri–Squared Test Taxonomy of Trichotomous Outcome Variables [$b_1 — b_3$] Examples of Repeated Measures Terminology

A Taxonomy of Tri–Squared Test Terminology: For the Creation of Inventive Investigative Instruments			
	Trichotomous Categorical Variables: As Indicators		
Trichotomous Outcome Variables	Discipline and Related Content Area:	Discipline and Related Content Area:	Discipline and Related Content Area:
	Awareness: *State of Existence*	Awareness: *State of Perception*	Awareness: *State of Experience*
b_1	Existence	Reality	Perspective
b_2	Non–Existence	Virtual	Non–Experienced
b_3	Dissipation	Eternal	Uninterested

The abovementioned is expressed as a series of Triostatistics Triangular Equation Models or "[TEM]s" using the Trichometry 3-4-5-6 GURT in the following graphical models:

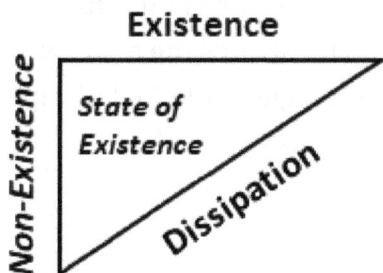

Existence

Non-Existence

State of Existence

Dissipation

Reality

State of
Perception

Virtual

Eternal

Perspective

State of
Experience

Non-Experienced

Uninterested

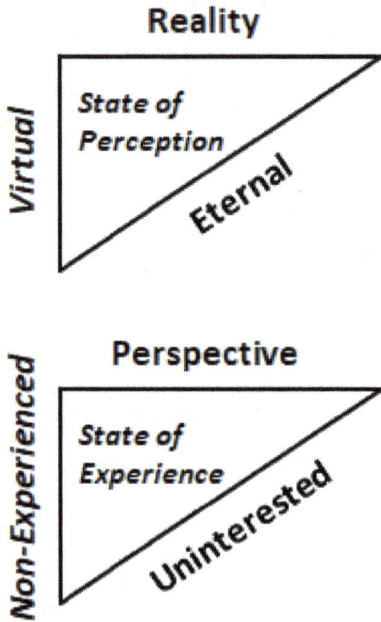

Table 28 attests to the series of Trichotomous Relations that can be used to qualitatively measure as a series of Trichotomous Categorical Variables in the holistic discipline of Awareness (in three different examples). The series of Trichotomous Outcome Variables provides (in units of three) the threefold possible "Trichotomous selections" that are the differentiated responses according to the mathematical Law of Trichotomy [as the set of Trichotomous Outcome Variables: b_1; b_2; or b_3]. Table 29 follows and illustrates the next series of possible repeated measures in a series of a single individual discipline.

Table 29
The Tri–Squared Test Taxonomy of Trichotomous Outcome Variables [b_1— b_3] Examples of Repeated Measures Terminology

	Trichotomous Categorical Variables: As Indicators		
A Taxonomy of Tri–Squared Test Terminology: **For the Creation of Inventive Investigative Instruments**			
Trichotomous Outcome Variables	Discipline and Related Content Area: Awareness: *State of Interaction*	Discipline and Related Content Area: Awareness: *State of Activity*	Discipline and Related Content Area: Awareness: *Elements of Being*
b_1	Empathy	Powerful	Internal
b_2	Aversion	Powerless	External
b_3	Uncaring	Empowered	Asternal

The abovementioned is expressed as a series of Triostatistics Triangular Equation Models or "[TEM]s" using the Trichometry 3-4-5-6 GURT in the following graphical models:

Empathy

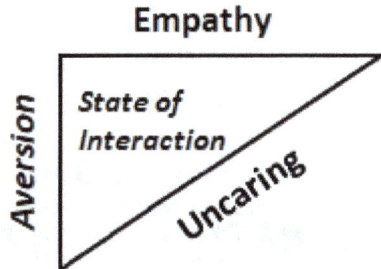

Powerful

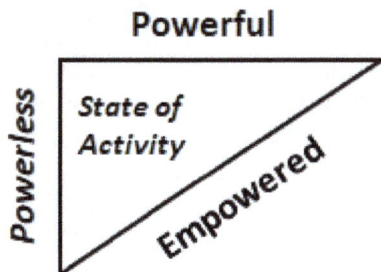

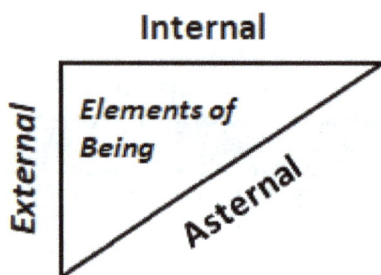

Internal

Elements of Being

External

Asternal

Table 29 substantiates the series of Trichotomous Relations that can be used to qualitatively measure as a series of Trichotomous Categorical Variables in the holistic discipline of Awareness (in three different examples). The series of Trichotomous Outcome Variables provides (in units of three) the threefold possible "Trichotomous selections" that are the differentiated responses according to the mathematical Law of Trichotomy [as the set of Trichotomous Outcome Variables: b_1; b_2; or b_3]. Table 30 follows and illustrates the next series of possible repeated measures in a series of two different disciplines.

Table 30

The Tri–Squared Test Taxonomy of Trichotomous Outcome Variables [b_1— b_3] Examples of Repeated Measures Terminology

	A Taxonomy of Tri–Squared Test Terminology: For the Creation of Inventive Investigative Instruments		
Trichotomous Outcome Variables	Trichotomous Categorical Variables: As Indicators		
	Discipline and Related Content Area:	Discipline and Related Content Area:	Discipline and Related Content Area:
	Measurement: *Operational Activity*	Measurement: *Activity Definitions*	Awareness: *Ethical Awareness*
b_1	Start	Begin	Give
b_2	Stop	End	Take
b_3	Pause	Away	Receive

The aforementioned is expressed as a series of Triostatistics Triangular Equation Models or "[TEM]s" using the Trichometry 3-4-5-6 GURT in the following graphical models:

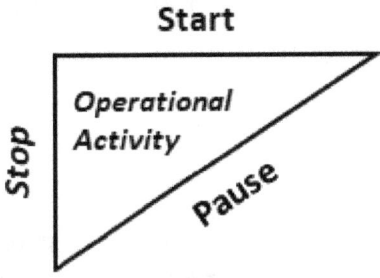

Start

Operational
Activity

Stop

Pause

Begin

Activity
Definitions

End

Away

Give

Ethical
Awareness

Take

Receive

Table 30 validates the series of Trichotomous Relations that can be used to qualitatively measure as a series of Trichotomous Categorical Variables in the comprehensive scientific discipline of Measurement and the holistic discipline of Awareness (in three different examples).

The series of Trichotomous Outcome Variables provides (in units of three) the threefold possible "Trichotomous selections" that are the differentiated responses according to the mathematical Law of Trichotomy [as the set of Trichotomous Outcome Variables: b_1; b_2; or b_3]. Table 31 follows and illustrates the next series of possible repeated measures in a series of three different disciplines.

Table 31
The Tri–Squared Test Taxonomy of Trichotomous Outcome Variables [b_1— b_3] Examples of Repeated Measures Terminology

	A Taxonomy of Tri–Squared Test Terminology: For the Creation of Inventive Investigative Instruments		
	Trichotomous Categorical Variables: As Indicators		
Trichotomous Outcome Variables	Discipline and Related Content Area:	Discipline and Related Content Area:	Discipline and Related Content Area:
	Existence: *State of Functionality*	Maturation: *State of Growth*	Philosophy: *Principles of Receiving*
b_1	Autonomy	Dependent	Earned
b_2	Dependence	Independent	Stolen
b_3	Emancipation	Self-Sufficient	Given

The abovementioned is expressed as a series of Triostatistics Triangular Equation Models or "[TEM]s" using the Trichometry 3-4-5-6 GURT in the following graphical models:

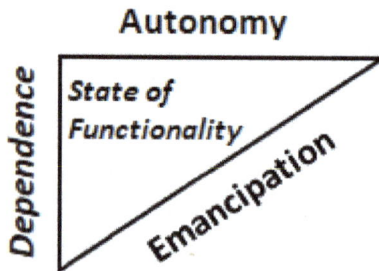

Autonomy

State of Functionality

Dependence

Emancipation

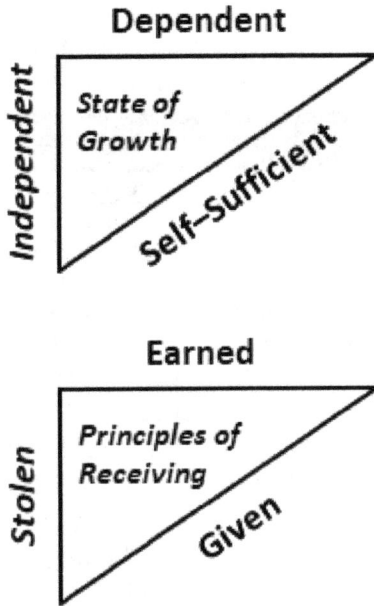

Dependent

Independent

State of
Growth

Self-Sufficient

Earned

Stolen

Principles of
Receiving

Given

Table 31 authenticates the series of Trichotomous Relations that can be used to qualitatively measure as a series of Trichotomous Categorical Variables in the holistic discipline of Existence, the universal and natural discipline of Maturation, and the academic discipline of Philosophy (in three different examples). The series of Trichotomous Outcome Variables provides (in units of three) the threefold possible "Trichotomous selections" that are the differentiated responses according to the mathematical Law of Trichotomy [as the set of Trichotomous Outcome Variables: b_1; b_2; or b_3]. Table 32 follows and illustrates the next series of possible repeated measures in a series of three different disciplines.

Table 32

The Tri–Squared Test Taxonomy of Trichotomous Outcome Variables [b_1— b_3] Examples of Repeated Measures Terminology

A Taxonomy of Tri–Squared Test Terminology: For the Creation of Inventive Investigative Instruments			
Trichotomous Outcome Variables	Trichotomous Categorical Variables: As Indicators		
	Discipline and Related Content Area: Psychology: *Structure of the Psyche*	Discipline and Related Content Area: Awareness: *State of Will*	Discipline and Related Content Area: Philosophy: *Morality and Ethics*
b_1	Superego	Committed	Conditions
b_2	Ego	Distracted	Disorder
b_3	Id	Non–Responsive	Random

The abovementioned is expressed as a series of Triostatistics Triangular Equation Models or "[TEM]s" using the Trichometry 3-4-5-6 GURT in the following graphical models:

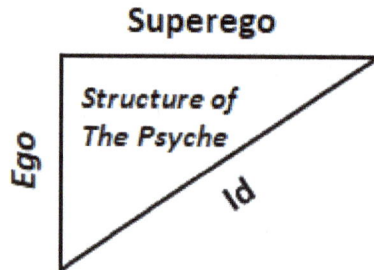

Superego

Structure of The Psyche

Ego

Id

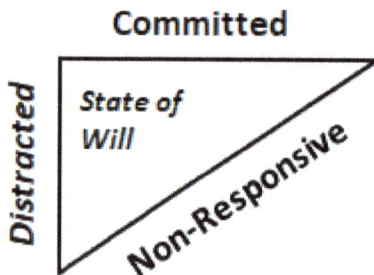

Committed

State of Will

Distracted

Non-Responsive

Conditions

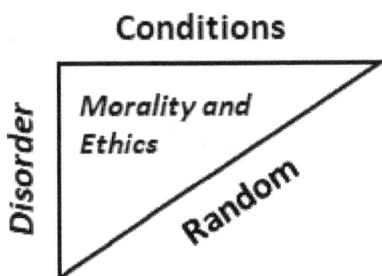

Table 32 confirms the series of Trichotomous Relations that can be used to qualitatively measure as a series of Trichotomous Categorical Variables in the scientific and academic discipline of Psychology, the holistic discipline of Awareness, and the academic discipline of Philosophy (in three different examples). The series of Trichotomous Outcome Variables provides (in units of three) the threefold possible "Trichotomous selections" that are the differentiated responses according to the mathematical Law of Trichotomy [as the set of Trichotomous Outcome Variables: b_1; b_2; or b_3]. Table 33 follows and illustrates the next series of possible repeated measures in a series of one individual discipline.

Table 33
The Tri–Squared Test Taxonomy of Trichotomous Outcome Variables [b_1— b_3] Examples of Repeated Measures Terminology

A Taxonomy of Tri–Squared Test Terminology: For the Creation of Inventive Investigative Instruments			
	Trichotomous Categorical Variables: As Indicators		
Trichotomous Outcome Variables	Discipline and Related Content Area:	Discipline and Related Content Area:	Discipline and Related Content Area:
	Measurement: *State of Structure*	Measurement: *Evidence of Structure*	Measurement: *Type of Structure*
b_1	Organized	Place	Pattern
b_2	Random	Remove	Non-Structure
b_3	Empty	Elsewhere	Nonexistence

The aforementioned is expressed as a series of Triostatistics Triangular Equation Models or "[TEM]s" using the Trichometry 3-4-5-6 GURT in the following graphical models:

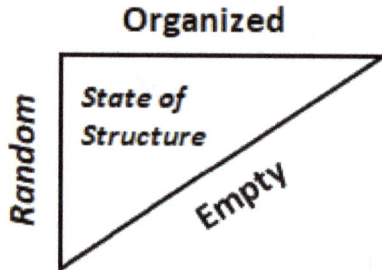

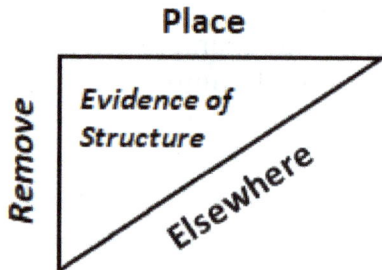

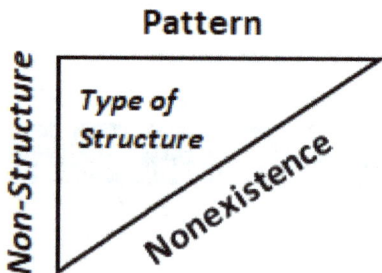

Organized

Random | *State of Structure* | *Empty*

Place

Remove | *Evidence of Structure* | *Elsewhere*

Pattern

Non-Structure | *Type of Structure* | *Nonexistence*

Table 33 supports the series of Trichotomous Relations that can be used to qualitatively measure as a series of Trichotomous Categorical Variables in the comprehensive scientific discipline of Measurement (in three different examples).

The series of Trichotomous Outcome Variables provides (in units of three) the threefold possible "Trichotomous selections" that are the differentiated responses according to the mathematical Law of Trichotomy [as the set of Trichotomous Outcome Variables: b_1; b_2; or b_3]. Table 34 follows and illustrates the next series of possible repeated measures in a series of three different disciplines.

Table 34
The Tri–Squared Test Taxonomy of Trichotomous Outcome Variables [b_1— b_3] Examples of Repeated Measures Terminology

	A Taxonomy of Tri–Squared Test Terminology: For the Creation of Inventive Investigative Instruments		
	Trichotomous Categorical Variables: As Indicators		
Trichotomous Outcome Variables	Discipline and Related Content Area:	Discipline and Related Content Area:	Discipline and Related Content Area:
	Measurement: *Activity Self–Assessment*	Awareness: *Involvement Compass*	Mathematics: *Primary Geometric Shapes*
b_1	Purpose	Concerned	Square
b_2	No Direction	Unconcerned	Circle
b_3	Not Involved	Disinterested	Triangle

The abovementioned is expressed as a series of Triostatistics Triangular Equation Models or "[TEM]s" using the Trichometry 3-4-5-6 GURT in the following graphical models:

Purpose / No Direction / Activity Self-Assessment / Not Involved

Concerned

Unconcerned | Activity Compass | Disinterested

Square

Circle | Primary Geometric Shapes | Triangle

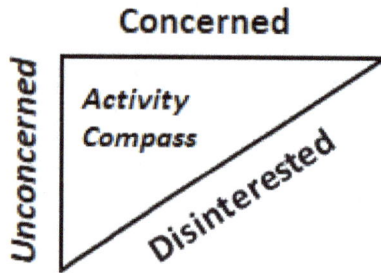

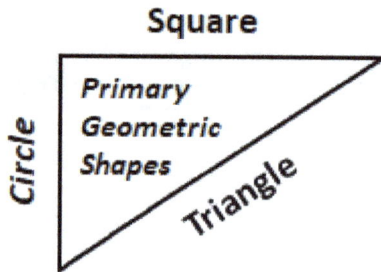

Table 34 determines the series of Trichotomous Relations that can be used to qualitatively measure as a series of Trichotomous Categorical Variables in the comprehensive scientific discipline of Measurement (in three different examples). The series of Trichotomous Outcome Variables provides (in units of three) the threefold possible "Trichotomous selections" that are the differentiated responses according to the mathematical Law of Trichotomy [as the set of Trichotomous Outcome Variables: b_1; b_2; or b_3]. Table 35 follows and illustrates the next series of possible repeated measures in a series of three different disciplines.

Table 35
The Tri–Squared Test Taxonomy of Trichotomous Outcome Variables [b_1— b_3] Examples of Repeated Measures Terminology

A Taxonomy of Tri–Squared Test Terminology: For the Creation of Inventive Investigative Instruments			
Trichotomous Outcome Variables	Trichotomous Categorical Variables: As Indicators		
	Discipline and Related Content Area:	Discipline and Related Content Area:	Discipline and Related Content Area:
	Science: *Components of the Earth*	Physics: *Elements of Nature*	Consumer Science: *Taste Identification*
b_1	Vegetable	Matter	Flavor
b_2	Mineral	Energy	Bland
b_3	Gas	Space	No Taste

The abovementioned is expressed as a series of Triostatistics Triangular Equation Models or "[TEM]s" using the Trichometry 3-4-5-6 GURT in the following graphical models:

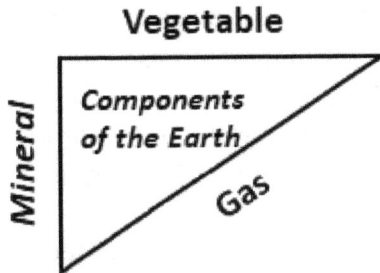

Vegetable

Mineral

Components of the Earth

Gas

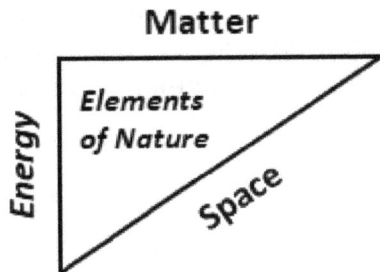

Matter

Energy

Elements of Nature

Space

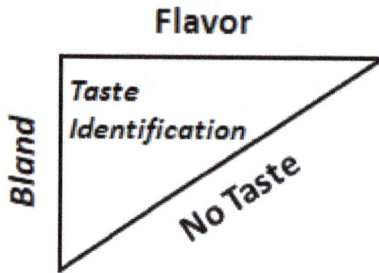

Table 35 is the final Table in this comprehensive list that establishes the last series of Trichotomous Relations that can be used to qualitatively measure as a series of Trichotomous Categorical Variables in the comprehensive scientific discipline of Measurement (in three different examples). The series of Trichotomous Outcome Variables provides (in units of three) the threefold possible "Trichotomous selections" that are the differentiated responses according to the mathematical Law of Trichotomy [as the set of Trichotomous Outcome Variables: b_1; b_2; or b_3]. The series of Tables 1 through 35 thus presented provide an all–inclusive set of examples Trichotomous repeated Measures data that can be used to create Tri–Squared Inventive Investigative Instruments for in–depth trichotomous inquiry. "A Differentiation of Three" is the key when it comes to the Law of Trichotomy and this is illustrated in the respective sets of Trichotomous Outcome Variables represented in each Table as: b_1; b_2; or b_3. This completes the "Triological Science of Trichometry" thirty-five Tables illustrating the "Trichotomous Tri–Values" found utilizing the Triostatistics Tri–Squared Test TOV and TCV within the framework of the Triostatistics "[TEM]" (Triangular Equation Modeling).

Chapter Seven follows and describes Triometry and Visualus.

Casting all your care upon him; for he careth for you.

I Peter 5: 7

Using Visualus Triometrically to Create a Solution using the Sequential Visualus Volumetrics Methodology

Using Visualus in terms of Trichometry as the mathematics of the Perceptology–Based Visualus Visioneering Volumetric is used as a Visualus Mathematical Analytic Engine to define the Educational Enterprise Equation (using the traditional methodology of Visualus) in the following sequential manner: (where, $9ab$ = 2 times the Trichometry 3-4-5-6 Golden Upright Right Triangle).

The Total Transitive Theorem of Visualus™ ©

$$\overset{a}{\underset{b\ \ c}{\text{T}}} \equiv \overset{abc}{\text{T}} \equiv \overset{a}{\underset{b\ \ c}{\text{∐}}}$$

"The Total Transitive Tandem of abc"

$$\overset{a}{\underset{b\ \ c}{\text{∐}}} \equiv 9ab \equiv \boxed{} \equiv$$

"The Total Transitive Trimetric of *abc*"

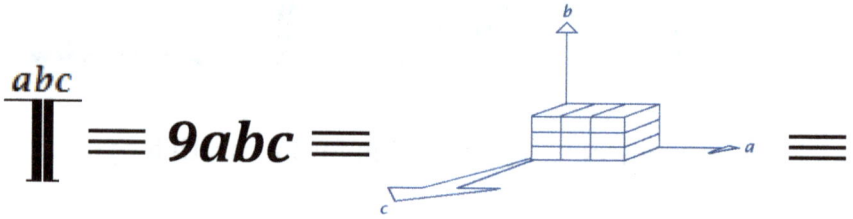

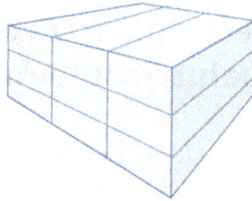

$$\dfrac{abc}{\rule{0pt}{0pt}\mathrm{I\!I}} \equiv 9abc \equiv$$

"The Total Transitive Trigma of *abc* into the Trigmatic Function"

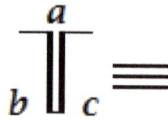

$$\overline{\underset{b\ \ \mathrm{I\!I}\ \ c}{a}} \equiv$$

216

TRICHOMETRY © *The Study of the Geometrics of the 3-4-5-6 Golden Upright Right Triangle in Cartesian Coordinates.* Osler Studios Incorporated ™ © Copyright 2022 All Rights Reserved.

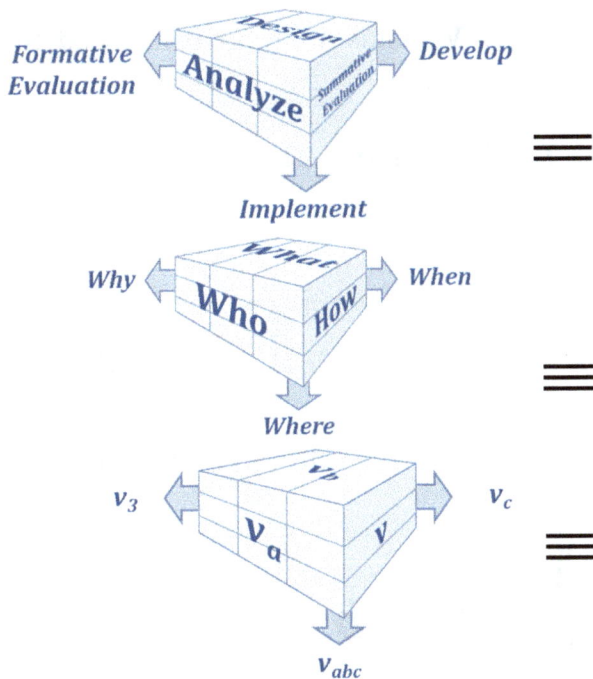

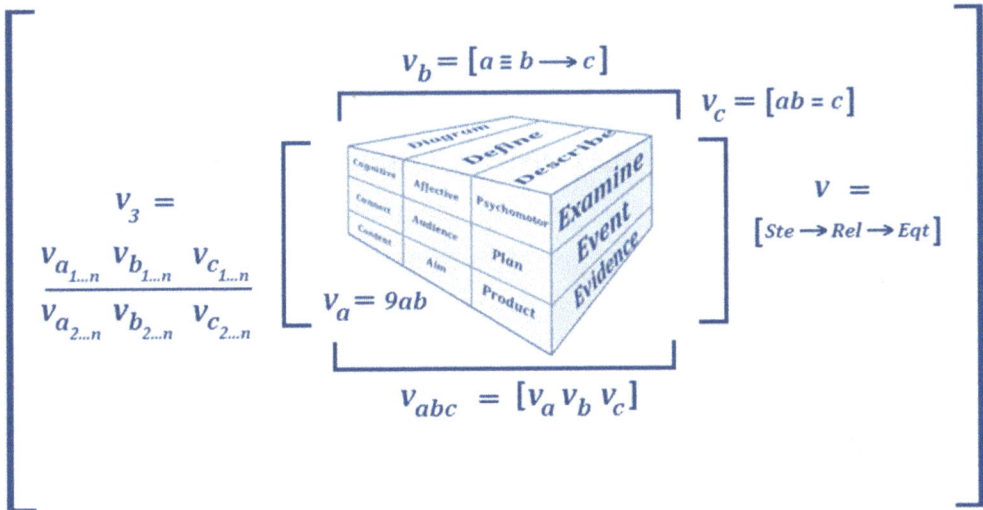

TRICHOMETRY ™ *The Study of the Geometrics of the 3-4-5-6 Golden Upright Right Triangle in Cartesian Coordinates.* Osler Studios Incorporated ™ © Copyright 2022 All Rights Reserved.

This creation of this Visualus solution transpired in the following manner:

The Total Transitive Theorem of Visualus™ ©

$$\underset{b}{\overset{a}{\text{╥}}}_{c} \equiv \overset{abc}{\text{╥}} \equiv \underset{b}{\overset{a}{\text{╨}}}_{c}$$

"The Total Transitive Tandem of *abc*"

$$\underset{b}{\overset{a}{\text{╨}}}_{c} \equiv 9ab \equiv \boxed{} \equiv$$

"The Total Transitive Trimetric of *abc*"

$$\overset{abc}{\text{╥}} \equiv 9abc \equiv$$

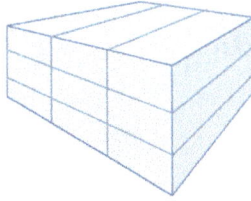

"The Total Transitive Trigma of *abc* into the Trigmatic Function"

$$\frac{a}{b \ \| \ c} \ =\!\!=$$

$$\equiv$$

$$\equiv$$

Formative Evaluation ← Analyze (Design, Summative Evaluation) → Develop

Implement

Why ← Who (What, How) → When

Where

v_3 ← v_a (v_b) v → v_c

v_{abc}

$$v_b = [\,a \equiv b \longrightarrow c\,]$$

$$v_c = [\,ab = c\,]$$

$$v_3 =$$

$$\frac{v_{a_{1\ldots n}} \; v_{b_{1\ldots n}} \; v_{c_{1\ldots n}}}{v_{a_{2\ldots n}} \; v_{b_{2\ldots n}} \; v_{c_{2\ldots n}}}$$

$$v_a = 9ab$$

$$v = [\,Ste \longrightarrow Rel \longrightarrow Eqt\,]$$

$$v_{abc} = [\,v_a \; v_b \; v_c\,]$$

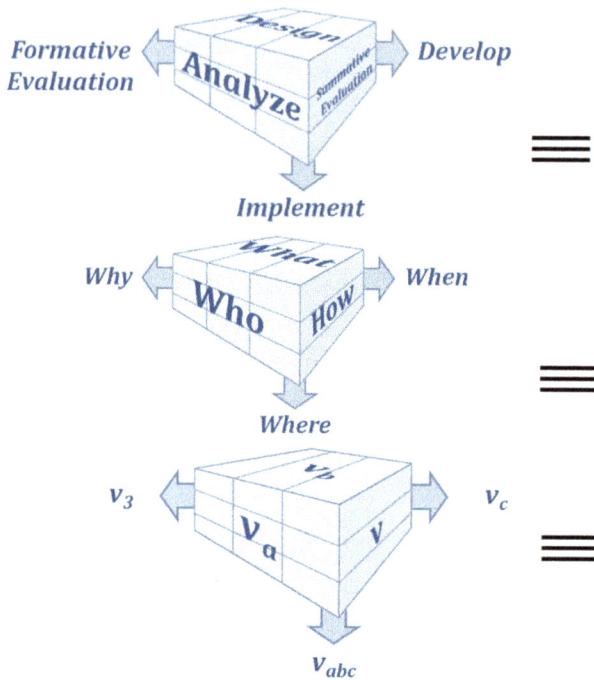

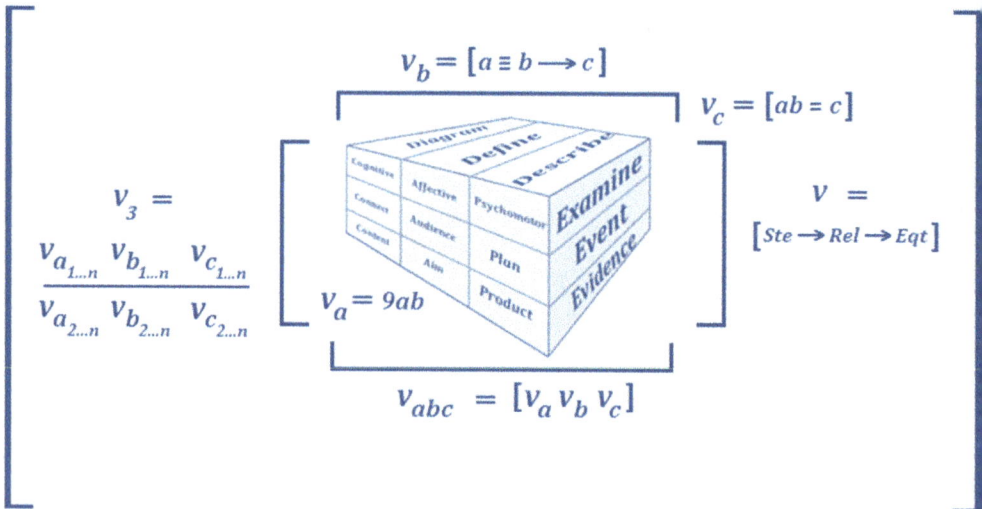

Chapter Eight follows and defines Trichometry in terms of the mathematical discipline of Triometry and its inherent "Triometrics".

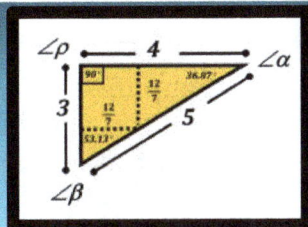

And the rain descended, and the floods came, and the winds blew, and beat upon that house; and it fell not: for it was founded upon a rock.

Matthew 7: 25

Defining the Mathematics of Triometry ©

The term "Triometry" is made up of the threefold terms: **"Tri"** (meaning "three" which stands for "Three-Coordinate" or "Tri-Coordinate"); **"O"** (which stands for "Order" or "Organization"); and **"Metry"** (which means "Measure"). Thus, the three combined terms create the following definition(s): **"Three-coordinate Order Measurement"** or **"Three-coordinate Organizational Measurement"**.

Triometry "Triometrics" as the Mathematics of Three-Coordinate (i.e., Tricoordinate) Organizational Structures

Triometry is the branch of mathematics that is the study of three-coordinate order, three-coordinate organizational methods, and the mathematical measures that carefully explain the three-coordinate organizational process. Triometry in terms of digital or "Interelectronic" computerized data is termed **"Ergonomic Techtonics"**. The term "Ergonomics" means "Usable" and "Techtonics" means "The Science or art of assembling, shaping, and constructing technology". Thus, Triometry as a mathematical methodology is the interactive in–depth investigation into the **"Ergonomic Structuring"** of computerized data for immediate access and archival purposes.

Triometry simplified is "Three-coordinate Organizational Measures" or "the ergonomic organization of an identified volumetric space in three dimensions for access and archival purposes".

The Triometry Methodology

The foundational methodology of Triometry is grounded in the science of **Perspectology**. Perspectology is the science of attitudinal redirection. It is the beneficial change in the Affective Domain of Learning (this is the Learning Domain that refers to attitudes and opinions) based upon verity and grounded in **"Veragogy"** (the art or science of teaching truth or verity). Perspectology in terms of **Triometry** is the science of positive perception in three dimensions. In terms of human behavior, Perspectology is concerned with the change of aberrant behavior into beneficial behavior completely grounded upon verity.

In terms of **Infometrics, Innovata**, and **Triometry** Perspectology has its foundation in Perceptology (the "Science of Comprehension") and is the process of organizing data three-coordinately according to a hierarchy of six structural methods that are categorized as: **Pack, Path, Parallel, Position, Purpose, and Portfolio**. Perception and perspective are critical elements in Triometry. Each is used to exemplify Triometrical mathematical techniques by grounding them in the respective sciences of Perceptology (the science of comprehension or understanding) and Perspectology (the science of attitudinal change). Each of the sciences provide the theoretical and conceptual framework that allows a "Triometrician" (a scientist who uses and is well versed in the mathematics of Triometry) to design and construct tangible beneficial solutions as "Interactive Data Structures". These three-coordinate "archivable" frameworks are creative interactive solutions created based upon need to empower a user by providing ready access to data.

Triometry Algorithms

Nesting Data as the Construction of
Triometric Interactive Ergonomic Data Structures

Triometric **Interactive Ergonomic Data Structures** is the branch of Triometry concerned with the method used to access electronic archival structures that repository various types of data. The creation of Interactive Ergonomic Data Structures is the "Nesting" of data into Interactive Data Organization structures. There are six hierarchal Interactive Ergonomic Data Structures in Triometry. They are six structural algorithms that are listed as: **(1) Pack; (2) Path; (3) Parallel; (4) Position; (5) Purpose; and (6) Portfolio**. A Pack is the least or lowest method of Triometric interactive data storage. Each preceding method of data storage that follows the Path method becomes larger, more detailed, and more complex in purpose, magnitude, and scale. Thus, the six data storage methods are defined and organized from lowest to highest in the following manner:

1.) **Pack** – A Start File that contains multiple files stored individually within it in a random unorganized manner ... towards ∞.

2.) **Path** – A Start File that is organized so that a file is located within a file within a file ... towards ∞.

3.) **Parallel** – A Start File that is organized so that it contains all files organized within a sub–file categorized by type within it.

4.) **Position** – A Start File that is organized so that it contains all files organized within multiple sub–files categorized by type containing multiple similar file types.

5.) **Purpose** – A Start File that is organized so that it contains all files organized within multiple sub–files categorized and organized by type or order of importance with each sub–file containing multiple similar file types.

6.) **Portfolio** – A Start File that is an organized data repository that contains all files organized within it organized into multiple sub–files at categorized and organized by type or order of importance containing an internal sub–file with multiple similar file types.

Visual and mathematical models of the six Interactive Ergonomic Data Structures are presented to aid in the understanding of how digital/electronic data is stored for archival and retrieval purposes.

Basic Triometric Algorithmic Symbols Used to Create Interactive Data Structures

= Indicates a Start File that is the primary containment structure used for Interactive Archival Data Storage.

I = Index = Three-coordinately this is = [Cuboid]

▲ = Data Type 1 = Three-coordinately this is = [Cone]

◻ = Data Type 2 = Three-coordinately this is = [Cylinder]

● = Data Type 3 = Three-coordinately this is = [Sphere]

In terms of three dimensions the data file types each represent a Cartesian Coordinate to provide each data type its own planar vector in the **Innovative Problem–Solving Model of Inventive Instructional Design**. This is represented in the following manner:

Where,

▲ = x = Three-coordinately this is = [Cone]

◻ = y = Three-coordinately this is = [Cylinder]

● = z = Three-coordinately this is = [Sphere]

The interactive three-coordinate data "Nesting" (the visual warehousing of information) within the **Visualus Isometric Cuboid** (which is used in Triometric data structures as the "archival volumetric space" as an outcome of the three data type Cartesian Coordinates) represents an ideal example of a "computerized nest" that serves as an interactive data storage repository. This exemplifies the archival of data within a Triometric organization structure (that can be accessed by a user). This is illustrated visually in the following model:

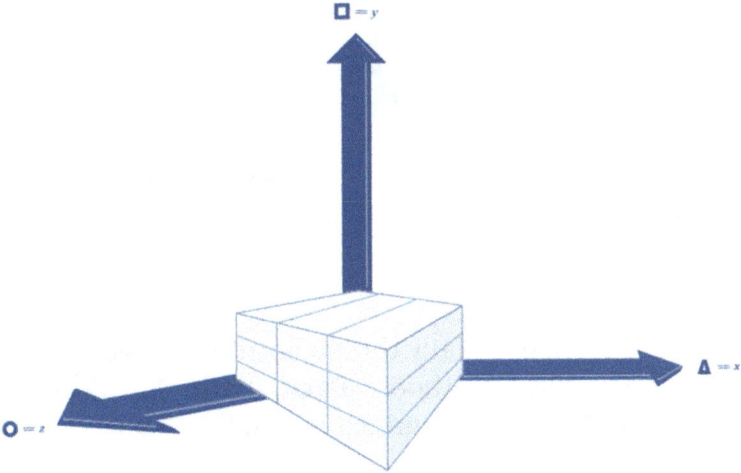

The "nested" data is three-coordinately archived along a specific categorical planar vector created by the Triometrician and housed within interactive data structure for retrieval and archival purposes. It is important to note that the three types of data are equal and interdependent (they are all individual and independent). Each data type is capable of being exchanged through their holistic collection and interactive connection within the Isometric Cuboid. This is illustrated in the following model:

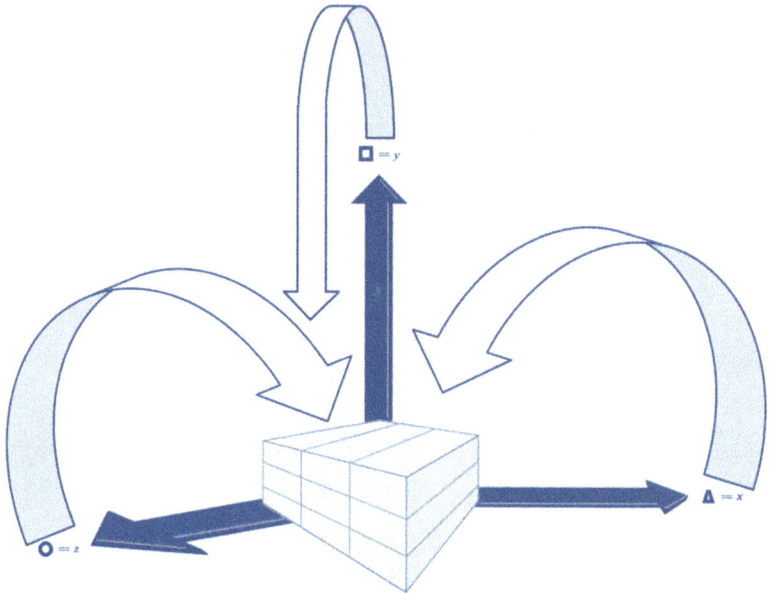

The process of organizing data three-coordinately according to the hierarchal six structural methods: **Pack, Path, Parallel, Position, Purpose, and Portfolio** are examined in detail in the next section.

Algorithmic Repository Structures

The following visual and mathematical hierarchal models are used to illustrate the six categorical Interactive Ergonomic Data Structures:

Pack

A Triometric Visual Model of a Pack Algorithm of Interactive Data Organization (this interactive data structure places data in an individual file):

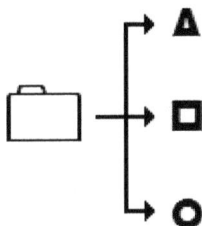

A Triometry Mathematical Model of a Pack Algorithm of Interactive Data Organization:

$$\mathbf{I} = \blacktriangle, \; \blacksquare, \; \bullet$$

Path

A Triometric Visual Model of a Path Algorithm of Interactive Data Organization (this interactive data structure is organized in a file according to file type):

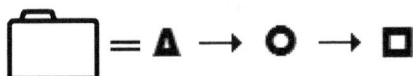

A Triometry Mathematical Model of a Path Algorithm of Interactive Data Organization:

$$I = \blacktriangle \Rightarrow \bullet \Rightarrow \square$$

Parallel

A Triometric Visual Model of a Parallel Algorithm of Interactive Data Organization (this interactive data structure is organized into sub–files according to type):

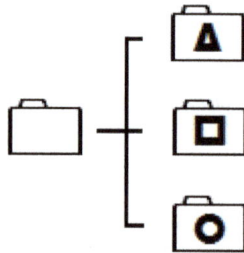

A Triometry Mathematical Model of a Parallel Algorithm of Interactive Data Organization:

$$I = \blacktriangle + \bullet + \square$$

Position

A Triometric Visual Model of a Position Algorithm of Interactive Data Organization (this interactive data structure is organized into multiple sub–files according to type):

$$\square \Big\{ \begin{array}{l} \boxed{\blacktriangle} = \blacktriangle + \blacktriangle + \blacktriangle \\ \boxed{\blacksquare} = \blacksquare + \blacksquare + \blacksquare \\ \boxed{\bullet} = \bullet + \bullet + \bullet \end{array}$$

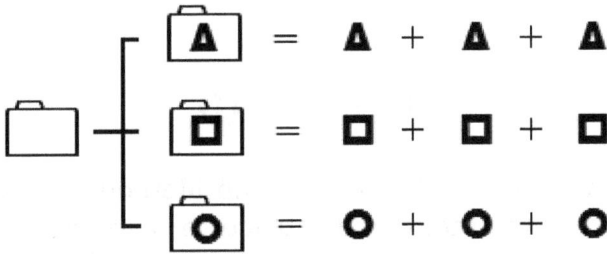

A Triometry Mathematical Model of a Position Algorithm of Interactive Data Organization:

$$I = \blacktriangle[n] + \bullet[n] + \blacksquare[n]$$

Purpose

A Triometric Visual Model of a Purpose Algorithm of Interactive Data Organization (this interactive data structure is organized into sub–files according to type by order of importance):

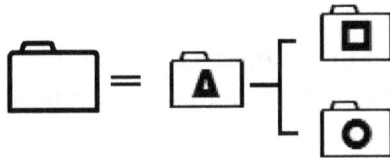

$$\square = \boxed{\blacktriangle} \Big\{ \begin{array}{l} \boxed{\blacksquare} \\ \boxed{\bullet} \end{array}$$

A Triometry Mathematical Model of a Purpose Algorithm of Interactive Data Organization:

$$\mathbf{I} = \mathbf{\Delta}\ \{\square, \mathbf{O}\}$$

Portfolio

A Triometric Visual Model of a Portfolio Method of Interactive Data Organization (this interactive data structure is organized into multiple sub–files according to type by order of importance):

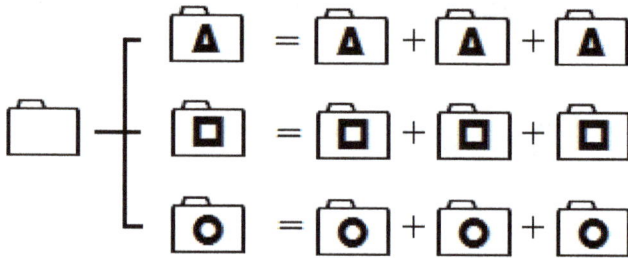

$$\boxed{\Delta} = \boxed{\Delta} + \boxed{\Delta} + \boxed{\Delta}$$

$$\boxed{\square} = \boxed{\square} + \boxed{\square} + \boxed{\square}$$

$$\boxed{\mathbf{O}} = \boxed{\mathbf{O}} + \boxed{\mathbf{O}} + \boxed{\mathbf{O}}$$

A Triometry Mathematical Model of a Portfolio Algorithm of Interactive Data Organization:

$$\mathbf{I} = \mathbf{\Delta}^{\infty} + \mathbf{O}^{\infty} + \square^{\infty}$$

Each of these models is interdependent and a Triometrician may use multiple Interactive Data Organizational methods to create the final comprehensive Data nest.

Chapter Nine follows and describes Trichometry in terms of trichotomous mathematical "concepts".

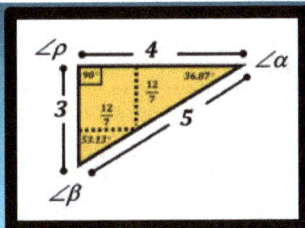

The Lord loveth the righteous.

Psalm 146: 8

Trichometry ©, Triometry ©, and Visualus ©

Triometry © is the mathematics of organizing content or "Mathematical Content Organization".

More broadly defined Triometry © is the mathematical term created by the author for a branch of mathematics that explores three-coordinate order by visually arranging content for archival purposes, ease of reference, and rapid delivery. "Triometry" in terms of "structure" is "the three-coordinate order or organization of data" or more specifically "the mathematics of the trimensional data arrangement". Triometry enhances Visualus by extending the rectilinear sequential Total Transitive Theorem of Visualus into an inverse explicative mathematical process through the Total Transitive Trimetric of *abc*.

The practical application of Triometry can be seen in the discipline that exemplifies the educational implications of Visualus called "Metacognetic Mechanics". The branch Metacognetic Mechanics called Technology Engineering illustrates Triometry in its use of data archival methods for instantaneous data retrieval in the creation of interactive electronic portfolios. Therefore, "Triometrical Technology Engineering" is the planning, design, development, construction, and production of "Interactive Digital E–Portfolios" that use triometric methods for hierarchal data archival for the specific purposes of data storage and retrieval.

The mathematics of Triometry work in tandem with the mathematics of Visualus (the two complement one another). Triometry expands Visualus to provide two operational methodologies in which Visualus can be actively applied to produce or define an innovative inventive solution. Visualus in turn provides Triometry with rationale and meaning as it mathematically defines and produces a tangible solution for Triometry as the archival methodology and creative archival solution exemplified in: "Standalone Interactive Electronic Portfolios".

Visualus © is the mathematics of visualizing and producing inventive solutions or "Mathematical Invention Visualization".

Examining Visualus through the Lens of Triometry: The Extension of Visualus into an Explicative Axiom of Problem–Solving

Triometry from a problem solution perspective provides **Visualus** (the innovative mathematics of invention visualization) greater weight and larger rationale pertaining to data organization methods.

In the metrics of Triometry, the mathematics of Visualus places great emphasis on explaining how a solution came into being as well as the methodology used to create that same problem–solving process as a solution. Thus, there are two arenas in which Visualus can be effectively utilized according to the organizational measures inherent to Triometry. The first arena is traditional or "sequential" in its approach and represents the traditional step–by–step problem–solving methodology used in Visualus. The second arena provides an "inverse" or "opposite" point of solution explanation that begins with the assumption that the solution has already been provided and Visualus is needed to explain and mathematically yield how the solution came into being and what was the process that led to its invention.

How each methodology is used greatly depends upon the "Visioneer" (an inventor/mathematician that uses Visualus) and their existing orientation to the problem that they are solving or have solved. In Visualus it is important to develop an understanding of the unique relationship and identical equivalence of the **"Trivium"** (Latin for "Three") which consist of the **"Total Tandem"**, the **"Total Trimetric"**, and the **"Total Trigma"**. All three of these elements are essential components of the Total Transitive Theorem of Visualus. Each is identical to the other but each has different meanings and differs in magnitude with different properties that are necessary to creating a Visualus solution.

The two Visualus "approaches" are illustrated mathematically through the Total Transitive Theorem of Visualus as follows:

Triometric Sequential Visualus:
Deriving a Solution Starting with the Problem by
Conducting an Extensive Examination to Create a Solution

$$\underset{b \quad c}{\overset{a}{\rm I\!I}} \equiv \overset{abc}{\rm I\!I} \equiv \underset{b \quad c}{\overset{a}{\rm I\!I}}$$

Triometric Inverse Visualus: Deriving a Final Solution Explaining the Invention Creation Process by Providing a Solution Mathematical Model

$$\overline{\underset{b}{\displaystyle\prod_{c}^{a}}} \equiv \overline{\underset{}{\displaystyle\prod}}^{abc} \equiv \underset{b}{\displaystyle\prod_{c}^{a}}$$

The essential element in both of these identities is the center element: "**The Total Trimetric of *abc***". The **Total Trimetric of *abc*** is boldly emphasized in the Total Transitive Theorem of Visualus in terms of content organizational mathematics or "Triometry". This bold emphasis indicates that the Trimetric mathematical identity within the formula is illustrating a capability of Visualus to be used in either a sequential or a inverse capacity.

This dual use of Visualus extends the Total Transitive Theorem into a "Total Transitive Axiom" or a "Totally Transitive Universally Accepted Truth" in terms of the Total Trimetric of *abc*. The Total Trimetric of *abc* now becomes the balancing point for either a sequential problem-solving or an explicative solution explanation point of view for the Visioneer. The identical [≡] forward $\left[\underset{b}{\prod_{c}^{a}} = \overline{\prod}^{abc} = \underset{b}{\prod_{c}^{a}} \right]$ and inverse $\left[\underset{b}{\prod_{c}^{a}} = \overline{\prod}^{abc} = \underset{b}{\prod_{c}^{a}} \right]$ structure of the Total Transitive Theorem of Visualus can now be examined organizationally as a sequential process that can be inversely explicative. The following mathematical model of the traditional Total Transitive Theorem of Visualus is paired with the inverse model underneath it:

$$\left.{}_b\right|\right|_c^{\,a} \;\equiv\; {}^{abc}\!\|\; \equiv\; \left|\right|_{b\;\;c}^{\quad a}$$

$$\left.{}_b\right|_c^{\,a} \;\equiv\; {}^{abc}\!\| \;\equiv\; {}_b\!\left|\right|\right._c^{\;a}$$

The Sequential and Inverse Visualus problem solution methods are the same. The only difference is the approach in which the Visioneer chooses to use Visualus based upon his or her perspective to explain, calculate, and model the problem–solving solution. Note the following:

In Sequential Visualus the Visioneer has not yet produced the solution. The problem is readily apparent however Visualus is needed to create a tangible invention and model the process in which the invention as solution came into being. The solution creation process then follows the six-vector sequence expounded by the **Innovative Problem–Solving Model of Inventive Instructional Design** on the Visualus Isometric Cuboid.

The six vectors each correspond to the six vectors of the Visualus Visioneering Volumetric Isometric Cuboid according to its surface area equation. This is illustrated as follows:

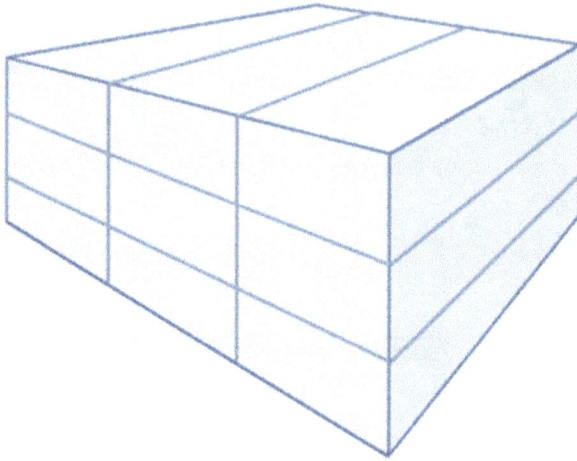

The Visualus Visioneering Volumetric Isometric Cuboid is divided into different Categorical Areas based upon the Innovative Problem–Solving Model of Inventive Instructional Design. The Categorical Areas are "Tricoordinately" divided into sub–areas based upon the volume of the Isometric Cuboid [*v = 9abc*] and the Surface Area of the Isometric Cuboid [*2(9ab + 3ac + 3bc)*]. This is illustrated in the next model as follows:

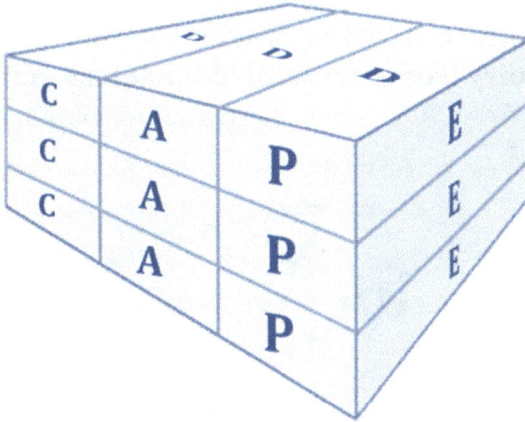

The Visualus Isometric Cuboid model and its meanings in terms of the Innovative Problem–Solving Model of Inventive Instructional Design can be illustrated visually in an **Explicative Innovative Inventive Instructional Design Isometric Cuboid Analytical Model** in the following manner:

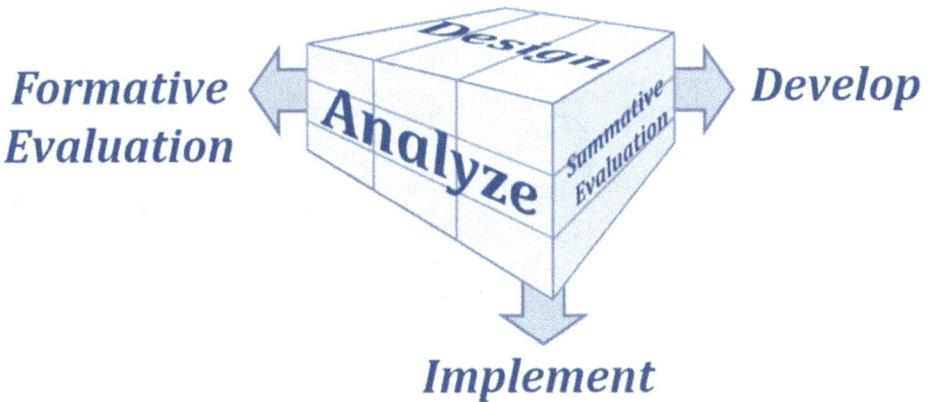

Tʀɪᴄʜᴏᴍᴇᴛʀʏ ™ © *The Study of the Geometrics of the 3-4-5-6 Golden Upright Right Triangle in Cartesian Coordinates.* Osler Studios Incorporated ™ © Copyright 2022 All Rights Reserved.

The above **Explicative Innovative Inventive Instructional Design Isometric Cuboid Analytical Model** can be simplified into an **Explicative Inquiry Isometric Cuboid Model** mean the following:

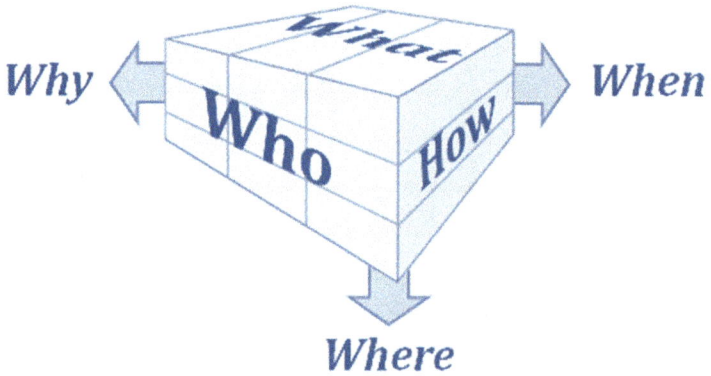

This can be visually illustrated graphically for purposes of clarity as follows:

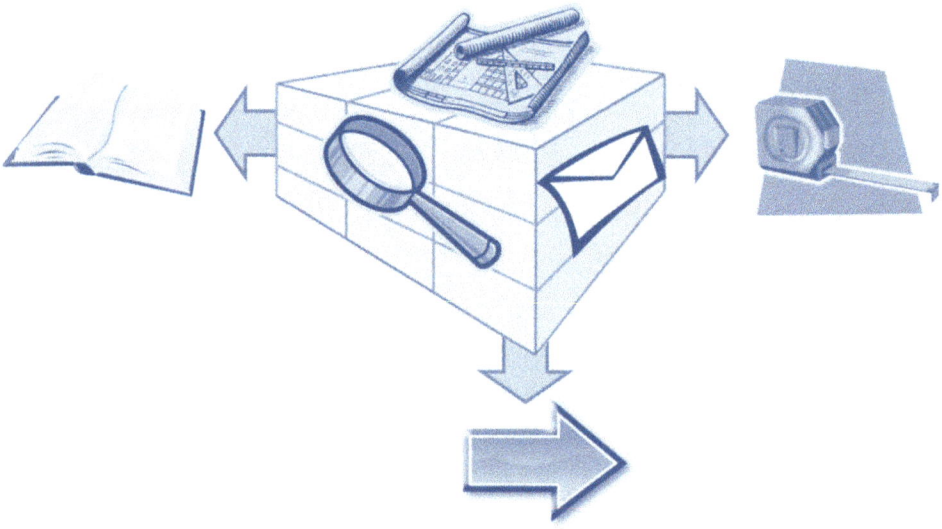

These visuals are further refined and defined to explain the sequential process of the Isometric Cuboid in the following manner:

Analyze Vector = Who = Determine = Magnifying Glass

Design Vector = What = Design = Blueprint

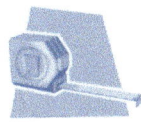

Develop Vector = When = Develop = Measurement Tape

Implement Vector = Where = Direction = Arrow

Formative Evaluation Vector = Why = Define = Direction

Summative Evaluation Vector = How = Deliver = Letter

These visual definitions more clearly articulate the meaning of Visualus in terms of the organizational methodology Triometry. This increases the scope of Visualus as the Theorem transitions into an Axiom written as:

$$\overset{a}{\underset{b \ \ c}{\top}} \equiv \overset{abc}{\top} \equiv \overset{a}{\underset{b \ \ c}{\parallel}}$$

The Sequential Visualus problem solution methodology can be represented rectilinearly in the following manner:

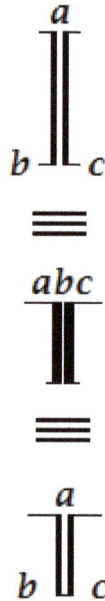

This particular orientation of the Triometric Inverse and Sequential Visualus problem solution model illustrates the Principles of Perceptological Increase that are illustrated mathematically. "Perceptology" is the "Science of Comprehension" or more specifically the "Science of Understanding". It is the ideal exemplification of Universal Instructional Design as the science of the "Innovative Problem–Solving Model of Inventive Instructional Design". The term comes from the Latin prefix "Percept" meaning "to understand" used in the Latin phrase "Percipio Percepi Perceptum" meaning "To Gain, To Learn, To Perceive". The foundational Latin term of "Percept" is used in the English term "Perception" meaning "Understanding". It is a theory that is concerned with the viability, marketability, and effectiveness of comprehending by means of interaction with forms of energy that allow beneficial or positive learning (the change in behavior as a result of acquiring of information).

Examples of beneficial forms of energy use to transmit information include but are not limited to: digital environments, distance learning environments, and electronic and digital tools. These tools include but are not limited to innovations in technology that enhance and make the learning environment more effective. The tools of Perceptology are: Visualus, Metacognetic Mechanics, Interactive Learning, and Interactive Technology. Thus, the structure of Perceptology is all–inclusive and umbrella–like building upon one another forming a "Perceptological Taxonomy" in the Ninefold Learning Domains in the following manner:

Perceptology = The Theoretical Framework; That forms a Field of Study through Universal Instructional design that creates solutions for increasing comprehension by transforming into the following —

Visualus = The Operational Mathematical Methodology; Producing mathematically modeled tangible solutions as innovative inventions in the learning environment by transforming into —

Metacognetic Mechanics = The Dynamic Discipline; Used to create learning solutions by transforming into —

Interactive Learning = The Conceptual Framework; that uses technology by transforming into —

Interactive Technology = The All–Inclusive Technology–Based Practice.

A scientist who gains and then shares comprehension through the study of verity is called a "Perceptologist". The term "Perceptological" describes a level comprehension that is developed and cultivated in the understanding of verity.

Perceptology is grounded in mathematical concepts and principles that greatly aid the "Perceptologist" (a problem–solving scientist who focuses on Perceptology) in the holistically describing the limitlessness of unity through understanding. There are five universal principles that best describe and illustrate this aspect of Perceptology they are: **Purpose, Structure, Magnitude, Expansion**, and **Abundance**. Each is respectively defined mathematically by an associated mathematical operation that clearly illustrates the principle in definitive terms. The mathematical operations are:

Addition = "Purpose" = Linear Increase;

Multiplication = "Structure" = Dilated Increase;

Exponentiation = "Magnitude" = Repetitive Increase;

Tetration = "Expansion" = Consistent Increase;

Extension (Immensity) = "Abundance" = Constant Increase

"Purpose" is defined in the "Science of Comprehension" as the first primary mathematical operation that is the simplest example of the two-coordinate unity of two separate amounts that increase through mutual collaborative cooperation. It is illustrated by the mathematical operation of **Addition** and is defined mathematically as follows:

$$a + a = 2a$$

"Structure" is defined in the "Science of Comprehension" as the two-coordinate increase in form as a result of the multiplication of unity that magnifies an established amount. It is illustrated by the mathematical operation of **Multiplication**. It is defined mathematically as follows:

$$a \times n = \underbrace{a + a + a + \ldots + a}_{n}$$

"**Magnitude**" is defined in the "Science of Comprehension" as the two-coordinate unity that dilates or grows a set amount from a base into a level of maturation as a result of the multiplicative increase of the set amount times itself. It is illustrated by the mathematical operation of **Exponentiation**. It is defined mathematically as follows:

$$a^n = \underbrace{a \times a \times a \times \ldots \times a}_{n}$$

"**Expansion**" is defined in the "Science of Comprehension" as a three-coordinate exponential increase that exponentially increases an exponent. Thus, it is an outward expansion that can continue infinitely or by a set amount. This illustrates the beginning of an possible manifestation of an outward increase without end. It is illustrated by the mathematical operation of **Tetration**. It is defined mathematically as follows:

$${}^{n}a = \underbrace{a^{a^{a^{\cdot^{\cdot^{a}}}}}}_{n}$$

"**Abundance**" is defined in the "Science of Comprehension" as the three-coordinate increase that infinitely expands an infinite endless eternal structure both outwardly and inwardly in all dimensions and in all directions. It is illustrated by the mathematical operation created by the author called "**Extension**" for "**Immensity**". Extension or to "mathematically extend" literally means "to promote growth or continuation" and ideally defines the mathematical concept of "Abundance". It is defined mathematically in two equal manners (with the exponent in the Symmetrical model clearly indicated by the statement written on the right side of the large base "a") as follows:

Diagonally as,

$$\begin{matrix} n \\ n \end{matrix} a \begin{matrix} n \\ n \end{matrix} = \begin{matrix} n...a \\ n...a \end{matrix} a \begin{matrix} a...n \\ a...n \end{matrix} =$$

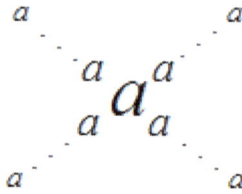

$$=$$

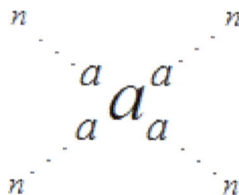

or

Symmetrically as,

$$n \, \overset{n}{\underset{n}{a}} \, n \;=\; a \cdots a \, \overset{\overset{a}{\vdots}}{\underset{\underset{a}{\vdots}}{a}} \, a \cdots a \;=$$

Where,

$$a \cdots a \, \overset{\overset{a}{\vdots}}{\underset{\underset{a}{\vdots}}{a}} \, a \cdots a$$

$$=$$

$$n \cdots a \, \overset{\overset{n}{\vdots}}{\underset{\underset{n}{\vdots}}{a}} \, a \cdots n$$

The "Extension" of Visualus in terms of Triometry, is format used to write the Total Transitive Theorem of Visualus in the following manner:

$$\overset{\displaystyle a}{\underset{\displaystyle b \; c}{\big\|}} \; \equiv \; \begin{matrix} \equiv \\[2pt] \overset{\displaystyle abc}{\underset{\displaystyle b \; c}{\big\|}} \\[2pt] \equiv \\[2pt] \overset{\displaystyle a}{\underset{\displaystyle b \; c}{\big\|}} \end{matrix} \; \equiv \; \overset{\displaystyle a}{\underset{\displaystyle b \; c}{\big\|}}$$

This illustrates that the identity is simultaneously rectilinear and illimited. Thus, Triometric Visualus identifies Visualus as a mathematical operation that has the dual properties of innovation and invention that are "free from limitation" or unlimited. This allows the Visioneer complete freedom in his or her creativity in the production of solutions.

At the core of this illimited innovative invention is the center element: **"The Total Trimetric of abc"**. As stated earlier, the "Total Trimetric of *abc*" is emphasized in the "Total Transitive Theorem of Visualus" in terms of Triometry. This emphasis indicates that the mathematical identity formula gives Visualus the capability to be used in both a sequential and inversed capacity.

The identical properties of elements that are contained in the Essential Theorem are unique to Visualus. This allows Visualus elements to be expressed in balanced and interesting ways. In terms of Triometry, the **"Infinite Visualus Mathematical Model"** illustrates that the Visualus elements are cyclical. This is possible because all of the elements in the identity are identical. **The Cyclical Model of the Infinite Visualus Mathematical Model** is illustrated as follows:

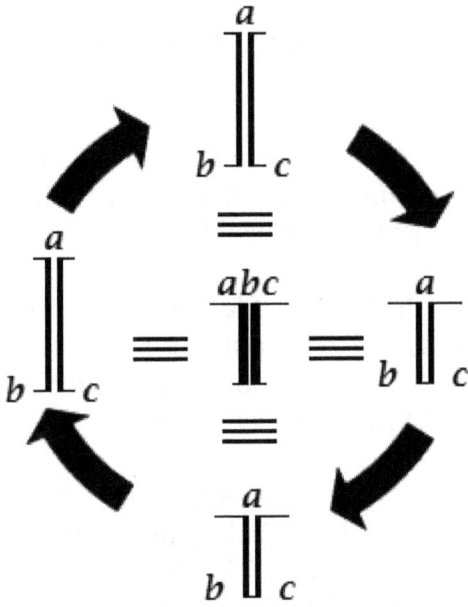

This concludes the book on the Triological Science of "Trichometry".

Previous Work by the Author that Formed the Foundation for TRICHOMETRY ™ ©

Osler, J. E. (1996). The Effects Of An Ergonomically Designed Computer–Based Tutorial On Elementary Students' Recall. Raleigh, NC: College of Education and Psychology – North Carolina State University.

Osler, J. E. (1996). A Mathematical Equation Expressing the Rectilinear Propulsion of the 100 and 110 Meter Hurdle Races ©. Durham, NC.

Osler, J. E. (2002). University Management Develop Program Comprehensive Report.

Osler, J. E. (2004). Dimensions of Teaching and Behavioral Impediments of Teaching Efficacy. Ideas About Teaching Efficacy: Sharing Perspectives. National Social Sciences Press. Gabe Keri.

Osler, J. E. (2004). The Crisis: Classroom Culture, Identifying and Analyzing Seven Factors That Disable An Effective Collegiate Teaching Methodology. A Long Way to Go: Conversations About Race By African American Faculty And Students. Peter Lang.

Osler, J. E. (2005) Technology Engineering: A Paradigm Shift In The Dynamics of Instruction; A New Philosophy of Education For Teaching In The Information Age. 2005 The South Atlantic Philosophy of Education Society Refereed 50th Conference.

Osler, J. E. (2005). Creating An Interactive Cognitive Economy: The Use of an Asynchronous Learning Network Course Management System to Develop an Interactive Community of Learners. Research Paper for 2005 The South Atlantic Philosophy of Education Society Refereed 50th Conference.

Osler, J. E. (2005). Technology Engineering: Developing, Implementing, and Infusing Interactive Metametric Learning Modules into an Asynchronous Learning Network to Develop an Interactive Community of Learners. Research Paper for The 2005 South Atlantic Philosophy of Education Society Refereed 50th Conference.

Osler, J. E. (2008). Σimply Σtatistics ©. A Comprehensive Guide to Statistical Formulae, Methodology, and Techniques. Durham, NC: Publishing Division, Osler Studios Incorporated™©.

Osler, J. E. (2008). THE VISION FINDER ©. Durham, NC: Publishing Division, Osler Studios Incorporated™©.

Osler, J. E. (2009). Σimply Σtatistics ©. The Handheld Edition. Durham, NC: Publishing Division, Osler Studios Incorporated™©.

Osler, J. E. (2009). The Osler Micro–Lending Strategy ©. Durham, NC: Research Division, Osler Studios Incorporated™©.

Osler, J. E. (2010). **INFOMETRICS** ™ ©. The Logistical Systemics and Strategic Practice of Empowering Learning through the Creation of an Ideal Learning Environment via Optimal Instruction. Durham, NC: Publishing Division, Osler Studios Incorporated™©.

Osler, J. E. (2010). **INNOVATA** ™ ©. The Science of Innovation and the Mathematics of Invention Expressed in the Discipline of Metacognetic Mechanics. Durham, NC: Publishing Division, Osler Studios Incorporated™©.

Osler, J. E. (2010). **INTERACTIVE INFORMATIVE INQUIRY** ™ ©: The Process of Investigating the Ergonomic Ideation of Technology Engineering and Optimal Instruction Durham, NC: Publishing Division, Osler Studios Incorporated™©.

TRICHOMETRY ̅©̅ *The Study of the Geometrics of the 3-4-5-6 Golden Upright Right Triangle in Cartesian Coordinates.* Osler Studios Incorporated ™ © Copyright 2022 All Rights Reserved.

Osler, J. E. (2010). **METAGRAPHIC USER INTERFACE DIGITAL LEARNING LABORATORY** ©: Interactive DVD. Durham, NC: Publishing Division, Osler Studios Incorporated™©.

Osler, J. E. (2010). **THE ADVANCED INTERACTIVE INTERELECTRONIC INVENTORY IDEATION LABORATORY** ©: Interactive DVD. Durham, NC: Publishing Division, Osler Studios Incorporated™©.

Osler, J. E. (2010). **THE VISION FINDER** ©. The Handheld Edition. Durham, NC: Publishing Division, Osler Studios Incorporated™©.

Osler, J. E. (2010). **VISUALUS** ™ © Visioneering Volumetrically: The Mathematics of the Innovative Problem–Solving Model of Inventive Instructional Design. Durham, NC: Publishing Division, Osler Studios Incorporated™©.

Osler, J. E. (2010). **PERCEPTOLOGY** ™ © The Science of Comprehension that is Universal Instructional Design thorough Visualus, Metacognetic Mechanics, and Optimal Instruction. Durham, NC: Publishing Division, Osler Studios Incorporated™©.

Osler, J. E. (2010). **ΣIMPLY ΣTATISTICS** ©. **THE HARDCOVER HANDHELD EDITION.** Durham, NC: Publishing Division, Osler Studios Incorporated™©.

Osler, J. E. (2010). **TECHTONICS** ™ ©. Optimal Learning via Instructional Solutions Developed through the Methodology of Technology Engineering. Durham, NC: Publishing Division, Osler Studios Incorporated™©.

Osler, J. E. (2010). **TRIMENSIONAL ANALYSIS** ™ ©. The Three-coordinate Analytics of Visualus Solution Metrics. Durham, NC: Publishing Division, Osler Studios Incorporated™©.

"I exhort therefore, that, first of all, supplications, prayers, intercessions, and giving of thanks, be made for all men; For kings, and for all that are in authority; that we may lead a quiet and peaceable life in all godliness and honesty. For this is good and acceptable in the sight of God our Saviour; Who will have all men to be saved, and to come unto the knowledge of the truth. For there is one God, and one mediator between God and men, the man Christ Jesus; Who gave himself a ransom for all, to be testified in due time. Whereunto I am ordained a preacher, and an apostle, I speak the truth in Christ, and lie not; a teacher of the Gentiles in faith and verity. I will therefore that men pray everywhere, lifting up holy hands, without wrath and doubting."

1ˢᵗ Timothy 2: 1–8

TRICHOMETRY ™ ©

TRICHOMETRY: The Study of the Geometrics of the 3-4-5-6 Golden Upright Right Triangle in Cartesian Coordinates is an informative guidebook developed to explain in detail the physical and mathematical operations that explain the 3-4-5-6 GURT.

A Final Dedication to Almighty GOD

At the completion of this book, I wish to once again acknowledge and thank Almighty GOD. May this book be a blessing to those who read it and may it aid them in the completion of each of their respective endeavors and make the world a better place as each person fulfills their divinely inherited purpose and destiny. In the blessed and holy name of my most precious Lord and Savior – Jesus Christ, AMEN.

Yours in Love, Truth, and Service

James E. Osler II, Ed.D.

"I thank God, whom I serve from my forefathers with pure conscience, that without ceasing I have remembrance of thee in my prayers night and day."

2nd Timothy 1:3

The following mathematical formulae and associated statements are affirmations related to Trichometry ©:

"My name is *Victory*."

[*v = 9abc*]

"Victory is New Birth All By to *Christ*"

and

[*v = 9xyz*]

"Victory is New Birth e*X*alting *Y*eshua onwards towards *Zion*"

"May *Almighty GOD* be glorified by each of us as we fulfill the purpose and destiny that he has placed in each of us."

"To *Almighty GOD* be the Glory!"

In *Jesus Christ's* Most Precious and Holy Name,

Amen.

ISBN: 978–0–9826748–17–5

TRICHOMETRY
THE GEOMETRICS OF THE 3-4-5-6 UPRIGHT RIGHT TRIANGLE

www.ingramcontent.com/pod-product-compliance
Lightning Source LLC
Chambersburg PA
CBHW060316100426
42812CB00003B/799